# Improving the Safety of Civilians

## A Protection Training Pack

# Oxfam GB

Oxfam GB, founded in 1942, is a development, humanitarian, and campaigning agency dedicated to finding lasting solutions to poverty and suffering around the world. Oxfam believes that every human being is entitled to a life of dignity and opportunity, and it works with others worldwide to make this become a reality.

From its base in Oxford in the United Kingdom, Oxfam GB publishes and distributes a wide range of resources for development and relief workers, researchers and campaigners, schools and colleges, and the general public, as part of its programme of advocacy, education, and communications.

Oxfam GB is a member of Oxfam International, a confederation of 13 agencies of diverse cultures and languages, which share a commitment to working for an end to injustice and poverty – both in long-term development work and at times of crisis.

For further information about Oxfam's publishing, and online ordering, visit www.oxfam.org.uk/publications

For information about Oxfam's development, advocacy, and humanitarian relief work around the world, visit www.oxfam.org.uk

# Acknowledgements

This training pack has been developed by Oxfam GB and aims to build the capacity of humanitarian field workers to improve the safety of civilians through humanitarian field programmes.

The materials were produced by Rachel Hastie, Sophia Swithern, Aine Bhreathnach, and Andrew Bonwick from Oxfam GB, based on original ideas by Jenny McAvoy. The materials are based on *Protection: An ALNAP Guide for Humanitarian Agencies* written by Hugo Slim and Andrew Bonwick, published by the Overseas Development Institute in 2005.

Many people have contributed towards this training pack and we would like to thank them for their input. Thanks to Awa Dabo from the United Nations Development Programme (UNDP) and Kate Pooler from the United Nations High Commissioner for Refugees (UNHCR) who provided support for the pilot work in Liberia, and to Ezekiel Davis, Caroline Bowah, Fatu Morris, and Kerien Palenah who helped facilitate the pilot workshops. We would also like to thank Vivien Walden, Kristen Geary, Christophe Beau, Sorcha O'Callaghan, Debbie Hunter, Simon Springett, Sophie Battas, Jo Nickolls, Claire Light, and Richard Nunn, all of whom have given very valuable comments and suggestions on the content.

# Improving the Safety of Civilians

## A Protection Training Pack

Oxfam

Practical Action Publishing Ltd
25 Albert Street, Rugby, CV21 2SD, Warwickshire, UK
www.practicalactionpublishing.com

First published by Oxfam GB in 2009
Reprinted by Practical Action Publishing

© Oxfam International 2007

Oxfam GB is registered as a charity in England and Wales (no. 202918) and Scotland
(SCO 039042).
Oxfam GB is a member of Oxfam International.

Paperback ISBN: 9780855986162
PDF ISBN: 9780855987749
Book DOI: https://doi.org/10.3362/9780855987749

A catalogue record for this publication is available from the British Library.

Cover: Mariamma Ali and Farah Sheik Omar leave Buseredo Supplementary Feeding
Centre, Ethiopia (2000). Crispin Hughes/Oxfam

Reasonable efforts have been made to publish reliable data and information, but the
author and publisher cannot assume responsibility for the validity of all materials or for
the consequences of their use.

The manufacturer's authorised representative in the EU for product safety is
Lightning Source France, 1 Av. Johannes Gutenberg, 78310 Maurepas, France.
compliance@lightningsource.fr

# Contents

**Introduction**    **7**

Using this training pack    **8**
Planning a workshop    **9**
Outline of training modules    **11**
Example agendas    **15**

## Opening the workshop

Pre-training questionnaire    18
Inspirational quotes    21
Inspirational quote cards    23
Introducing the training    27

## Module 1
### What is protection?

**Session 1: What is protection?**
Trainer's notes    30
Resource materials for trainers    30
Session plan    31
Role play briefing    32
Handout 1 – What is protection?    33
Templates for visual aids    34

**Session 2: Tools for protecting civilians in conflict**
Trainer's notes    38
Resource materials for trainers
Session plan    42
Handout 2 –International standards for civilian protection    44

**Session 3: Who protects?**
Trainer's notes    45
Resource materials for trainers    45
Session plan    46
'Who protects?' cards    47
Handout 3 – Who protects?    55

**Session 4: Vulnerability analysis**
Trainer's notes    56
Resource materials for trainers    56
Session plan    57
'Play the Role of' cards    58

## Module 2
### Planning a programme

**Session 1: Protection analysis**
Trainer's notes    68
Resource materials for trainers    68
Session plan    69
Protection analysis template    70
Mini case studies    71

**Session 2: Gathering and managing information**
Trainer's notes    72
Resource materials for trainers    72
Session plan    73
Role play briefings    74
Scenarios    76
Handout 4 – Gathering and managing information
for protection    77

**Session 3: Risk analysis**
Trainer's notes    80
Resource materials for trainers    80
Session plan    81
Template for feedback    82
Handout 5 – Managing risk    83

# Contents

## Module 3
### Mainstreaming protection

**Session 1: Mainstreaming protection**

| | |
|---|---|
| Trainer's notes | 88 |
| Resource materials for trainers | 88 |
| Session plan | 89 |
| Programme management cycle grid | 90 |
| Handout 6 – Mainstreaming protection | 91 |

**Session 2: Options for mainstreaming protection**

| | |
|---|---|
| Trainer's notes | 93 |
| Resource materials for trainers | 93 |
| Session plan | 94 |

**Group 1: Co-ordination with other agencies**

| | |
|---|---|
| Scenarios | 95 |
| Template for feedback | 96 |
| Handout 7 – Co-ordination and referral for humanitarian protection | 97 |

**Group 2: Adapting the programme**

| | |
|---|---|
| Scenarios | 99 |
| Template for feedback | 100 |
| Handout 8 – Adapting the programme to protect | 101 |

**Group 3: Local-level advocacy**

| | |
|---|---|
| Scenarios | 103 |
| Template for feedback | 104 |
| Handout 9 – Local-level advocacy for protection | 105 |

**Session 3: What would you do?**

| | |
|---|---|
| Trainer's notes | 108 |
| Resource materials for trainers | 108 |
| Session plan | 109 |
| 'What would you do' cards | 110 |

## Module 4
### Programming for protection

**Session 1: Options for protection programming**

| | |
|---|---|
| Trainer's notes | 117 |
| Resource materials for trainers | 117 |
| Session plan | 118 |

**Group 1: Sexual violence**

| | |
|---|---|
| Trainer's notes | 119 |
| Resource materials for trainers | 119 |
| Session plan | 120 |
| Template for feedback | 121 |
| Handout 10 – Sexual violence | 122 |
| Case study | 123 |
| Programming options cards | 124 |
| Alternative session | 125 |

**Group 2: Displacement**

| | |
|---|---|
| Trainer's notes | 126 |
| Resource materials for trainers | 126 |
| Session plan | 127 |
| Template for feedback | 128 |
| Handout 11 – Supporting durable solutions to displacement | 129 |
| Case study | 131 |
| Programming options cards | 132 |

**Session 2: Objective setting**

| | |
|---|---|
| Trainer's notes | 133 |
| Resource materials for trainers | 133 |
| Session plan | 134 |
| Template for feedback | 136 |
| Handout 12 – Terms used in objective setting | 137 |

Dave Clark/Oxfam

**Session 3: Indicators and monitoring**

Trainer's notes                                    138
Resource materials for trainers                    138
Session plan                                       139
Handout 13 – Indicators and monitoring             140

# Core exercises

Using the core exercises                           144
**True or false?**                                 145
Trainer's notes                                    145
Session plan                                       146
True or false statements                           147
**Agree or disagree**                              152
Trainer's notes                                    152
Session plan                                       153
Agree or disagree statements                       154
Agree or disagree cards                            159
**Protection stories**                             163
Resource materials for trainers                    163

# Closing the workshop

Dealing with outstanding issues                    170
Participants' contact details                      170
Feedback and evaluation                            170
Practicalities                                     171
Trainer's feedback                                 171
Follow-up                                          171
Feedback form                                      172

# Injustice anywhere is a threat to justice everywhere

Martin Luther King

# Introduction

In 1991 during the first Iraq war, a destitute little girl in the holy city of Safwan stood in the middle of a crowd of displaced people with a placard around her neck: 'We don't need food, we need safety.' She epitomised a shift in the understanding of what civilians expect from the international community and the role of humanitarian agencies in securing it. The aftermath of the Rwandan genocide in 1994 saw a further shift. The militarisation of the refugee camps in Zaire (now the Democratic Republic of Congo) led agencies to question whether their provision of aid in the camps contributed to the insecurity in the region and had actually undermined the safety of the people they were trying to assist.

These are the two sides to protection: first ensuring that humanitarian action does not expose civilians to further risks, and second, proactively using humanitarian action to improve the safety of civilians. Although the two can never be entirely separated, this training pack focuses on the latter: designing programmes that aim to improve the safety of civilians being subjected to violence, coercion, or deliberate deprivation.

International humanitarian, refugee, and human-rights law is an important backdrop to protection. It sets standards for the way people should be treated; it can help us locate responsibility for action or inaction; and it can sometimes be used to persuade the relevant authorities to take action to protect people. However, humanitarian workers do not need a detailed understanding of the law in order to design and deliver protection programmes. It is thus only dealt with very lightly in this training pack.

Similarly, while monitoring and reporting on 'violations' or educating people on their rights can play a part in protection programmes, they are not the only options available. This training pack sets out a range of activities that can be used to improve the safety of civilians. Some actions can be used to reduce the level of threat against civilians: advocacy to persuade others to protect; capacity-building to help them protect; and 'presence' to deter potential perpetrators. Others can be used to reduce people's vulnerability or exposure to threats: the provision of assistance or information; and helping civilians have a stronger 'voice' to negotiate or advocate for their own safety.

As abuses and violations targeted at civilians continue, and in some cases intensify, it is important for humanitarian actors to be prepared to take all measures they can to improve the safety of civilians. The aim of this training pack is to build the capacity of humanitarian practitioners to take very real and practical actions to improve the safety of civilians.

# Using this training pack

The training pack is set out as a three-day workshop, although all the materials can be used flexibly and the content adapted to specific groups or circumstances. These materials emphasise experiential learning – learning by doing. The methods used range from facilitator-led discussions, games, and role plays to small group exercises.

Module 1 introduces the idea of humanitarian protection in a practical sense, examines the roles of different protection actors, and looks at what standards exist to protect civilians and how they can be used. Module 2 focuses on gathering and managing information about protection, including doing a protection analysis. Module 3 focuses on mainstreaming protection in humanitarian programming, following the stages of the project cycle management. Module 4 looks at some programming options addressing sexual violence and durable solutions to displacement, leading on to the participants creating their own logframes for protection programmes to respond to the threats identified in Module 1.

The training pack also includes some core exercises, which are central to the whole workshop as they highlight issues of individual attitudes, beliefs, and prejudices, and look at what practical actions we can take in a range of situations.

We would like to encourage users of this training pack to try using the materials in different ways. Example agendas for a three-day and a one-day workshop are included at the end of this section. The core exercises are the most flexible tools and can be used in many different ways – suggestions are included in the trainer's notes for each exercise.

In order to deliver the training outlined in the manual you will need:
- knowledge and experience of human rights and humanitarian protection
- training and facilitation skills
- adequate preparation time for familiarisation with the materials, planning, photocopying, preparing flip charts, etc.

It is not necessary to have a legal qualification to use these materials, but some knowledge of the relevant bodies of law will be necessary as well as an understanding of programme planning, particularly project cycle management. In order to carry out a three-day workshop we recommend a team of two to three trainers with a balance of skills and experience in these areas.

The training is aimed at people who have little or no experience of humanitarian protection. The materials have been designed for use where there is no electricity or electronic equipment – so there is no need for lap tops, data projectors, and so on, but some materials do need photocopying in advance.

## This training pack consists of:

- **Trainer's manual** (the main book)

- **Cards and posters** (available in colour at the back of the book to pull out. Extra copies can be photocopied or printed from the CD.)

- **CD**: contains the full text of *Improving the Safety of Civilians*.

## Additional reading

Trainers using this pack are recommended to read *Protection: An ALNAP Guide for Humanitarian Agencies* by Hugo Slim and Andrew Bonwick, published by the Overseas Development Institute in 2005 (www.alnap.org/publications/protection/index.htm).

This pack draws on and complements other training materials and resources relating to protection. The following are particularly useful:

Reach Out's *Training Kit on Refugee Protection* (2005) provides comprehensive materials and resources on refugee protection and includes additional materials on gender-based violence and the protection of internally displaced people (www.icva.ch/doc00001528.html).

*The Sphere Training Package* produced by The Sphere Project (2004) contains a very useful module on The Humanitarian Charter that includes training materials on refugee, human-rights, and international humanitarian law (www.sphereproject.org).

*Action Against Small Arms: A Resource and Training Handbook* by Jim Cole and Henry Smith, produced by International Alert, Oxfam GB, and Saferworld (2003) contains very useful materials on developing an advocacy strategy and managing risk (publications.oxfam.org.uk/oxfam/display.asp?K=9780855984977).

*The Oxfam Gender Training Manual* by Suzanne Williams, produced by Oxfam (1994) contains many useful exercises and materials, including materials specifically focusing on sexual and gender-based violence (publications.oxfam.org.uk/oxfam/display.asp?K=9780855982676)

The IASC *Guidelines for Gender-based Violence Interventions in Humanitarian Settings* (2005) provide very useful definitions and action sheets (www.humanitarianinfo.org/iasc/content/subsidi/tf_gender/gbv.asp).

# Planning a workshop

## Participant numbers

The number of participants in each workshop will usually be 20–24. For smaller or larger groups the materials may need some adaptation, especially for small group work that has been designed for 5–6 participants per group.

## The training team

The workshop uses intensive methods and requires active participation by all trainers in order to steer groups and continually challenge them to develop their learning. Although these materials contain a lot of content, this will need to be supplemented by the training team's knowledge and experience.

For a group of 20–24 participants you will need at least two trainers. Trainers should ideally have complementary skills and knowledge – one might be a protection specialist, the other have experience of training in the relevant context and language.

The training team should agree the agenda in advance and decide who will lead each session. The person leading the session is responsible for doing any necessary background reading and ensuring that all materials are prepared for that session. In many cases the other trainers will be involved in a supporting role, particularly in small group work. The training team should meet after each day of the workshop to debrief and plan the following day, and at the end of the event to debrief and give feedback to each other.

## Preparation

Running the workshop requires considerable practical preparation (preparation of training materials, handouts, etc. – details are given in the trainer's notes for each session) as well as preparatory reading and familiarisation. Trainers should ensure they have prepared trainer's notes and that all handouts and flip charts have been copied in advance.

## Training venue

The venue should be large enough for participants to sit in plenary (in a semi-circle) facing a wall on which the trainer can stick the visual aids to create a 'talking wall'. There needs to be room for four tables to be set up for group work outside the semi-circle (more tables will be needed for larger groups). The vulnerability analysis exercise asks participants to line up along a wall with an open space in front of them. In setting up the room try to ensure that such a space could be made available without too much movement of furniture. If any participant has a mobility impairment all exercises involving movement should be adapted so that everyone can participate in the exercise in the same way.

## Equipment

Flip-chart markers, blu-tack, and a supply of flip-chart paper are required. A flip-chart stand would be useful.

## Pace and timings

Timings are provided for all sessions, but these are only guides. It is possible for the trainer to adapt each session to fit the time available and the level of the group. The core exercises have been designed to be used flexibly, so can be used to reinforce learning quickly or to stimulate lengthy discussion. In any workshop the trainer must be prepared to adapt the agenda to the specific needs of the participants who may wish to spend longer on an exercise or discussion than planned. The trainers need to judge how to move the workshop forward. If a group becomes sidetracked on an issue that cannot be resolved in the workshop it may be necessary to 'park' the issue by getting the basic details and deciding how it will be followed up, before moving back to the planned agenda.

Check local customs and practices – for example you may need to plan in extra time for prayers.

**Please note:** For more detailed guidance on planning and running a training event, you may find it useful to refer to The Sphere Training Package Trainer's Guide (see p. 8 under 'Additional reading') which contains comprehensive information on organising workshops and training events.

# Outline of training modules

| Module | Participants will: | Topics | Methods, timing, and resources |
|---|---|---|---|
| **Module 1: What is protection?**<br><br>**Maximum 3 hours** | Understand what humanitarian protection is, the main protection threats, the concept of a risk-based approach, and different modes of action<br><br>Be familiar with the international standards and principles for humanitarian protection<br><br>Understand the key players in protection and the roles they play in theory and reality<br><br>Understand the factors that make people vulnerable to threats and how to analyse power differences | **Session 1: What is protection?**<br><br>Outlines the key definitions and concepts used in protection | Role play followed by short interactive presentation using visual aids to build a 'talking wall' to help participants retain information and to help participants not working in their own language.<br><br>Handouts provided.<br><br>**20–30 mins** |
| | | **Session 2: Tools for protecting civilians in conflict**<br><br>An overview of the international standards for civilian protection and how they can be used by humanitarian organisations | Facilitator-led discussion followed by true or false game involving movement around the room which helps to maintain energy levels, to reinforce key learning points and stimulate discussion and debate.<br><br>Handouts provided.<br><br>**45 mins** |
| | | **Session 3: Who protects?**<br><br>Identifies the key players in humanitarian protection, what responsibilities, roles, or actions they might have/take, and how they function in reality | Group exercise using pre-prepared cards. Involves working in groups around a table. Can be followed by true or false game involving movement around the room to reinforce key learning points and stimulate debate.<br><br>Templates for cards included in training pack.<br><br>Handouts provided.<br><br>**30–60 mins** |
| | | **Session 4: Vulnerability analysis**<br><br>Participants identify the factors that make people vulnerable and develop an understanding of power differences | A group role play followed by a facilitator-led discussion with participants moving around the room. Participants need to line up against a wall with a clear space in front of them.<br><br>**40–60 mins** |

| Module | Participants will: | Topics | Methods, timing, and resources |
|---|---|---|---|
| **Module 2: Planning a programme**<br><br>**Maximum 3.5 hours** | Be able to carry out a basic protection analysis, identifying the threat, who is vulnerable, who are the perpetrators, who has responsibility for protection, and who is able to take action | **Session 1: Protection analysis**<br><br>Participants apply knowledge gained in Module 1 to analyse a specific protection scenario | Facilitator-led process in plenary to identify main threats followed by small group work using pre-set analysis formats. Each group feeds back in plenary.<br><br>**60 mins** |
| | Know the different ways of gathering and managing information, the practical and ethical issues involved, and understand best practice<br><br>Objectively understand the risks involved in gathering and responding to information about protection threats, how to mitigate these, and how a decision should be made on if/how to proceed | **Session 2: Gathering and managing information**<br><br>Participants identify some of the challenges, issues, and best practice in relation to gathering reliable information about protection and the sensitivities in handling it appropriately | Participants divide into groups to perform role plays and/or discuss scenarios, followed by discussion in plenary and a 'Do's and Don'ts' session where participants move around the room to post their recommendations. Handouts provided.<br><br>**90 mins** |
| | | **Session 3: Risk analysis**<br><br>Participants identify and analyse some of the potential risks of protection work to beneficiaries, staff, and organisations, and make recommendations on how to deal with them | Participants divide into two groups to analyse a developing scenario using a pre-set risk assessment format. Each group feeds back in plenary.<br><br>**60 mins** |

| Module | Participants will: | Topics | Methods, timing, and resources |
|---|---|---|---|
| **Module 3: Mainstreaming protection**<br><br>**Maximum 3.5 hours** | Be familiar with the programme cycle management model and the minimum requirements to mainstream protection at each stage<br><br>Understand different options for minimum response to protection threats<br><br>Be able to apply the different options to common scenarios | **Session 1: Mainstreaming protection**<br><br>An overview of the project cycle management leading participants to identify the key steps they would need to take to mainstream protection at each stage | Trainer-led presentation followed by work in four small groups and discussion in plenary.<br>Handouts provided.<br>**45–60 mins** |
| | | **Session 2: Options for mainstreaming protection**<br>**Group 1: Co-ordination with other agencies**<br>**Group 2: Adapting the programme**<br>**Group 3: Local-level advocacy**<br><br>Participants look at the practical application of each option, how to overcome some of the challenges, and identify best practice | Participants divide into three groups, each discussing scenarios related to a different topic and filling in pre-set templates. Each group feeds back to plenary.<br>Handouts provided.<br>**90 mins** |
| | | **Session 3: What would you do if?**<br><br>Participants apply what they have learnt in Sessions 1 and 2 to respond to specific scenarios | This session can be run in a number of ways. Participants choose scenarios from pre-prepared cards and explain what they would do in response.<br>**45 mins** |

| Module | Participants will: | Topics | Methods, timing, and resources |
|---|---|---|---|
| **Module 4: Programming for protection**<br><br>**Maximum 3.5 hours** | Understand the issues and principles associated with displacement and sexual violence, and programming options for addressing them<br><br>Be able to use a logical framework to design the objectives, outcomes, outputs, and activities of a protection programme to respond to a specific threat<br><br>Know how to set indicators for protection programming and monitor progress against them | **Session 1: Options for protection programming**<br>**Group 1: Sexual violence**<br>**Group 2: Displacement**<br>Participants analyse a case study and choose activities to form a programme in response | The participants divide into two groups that work simultaneously, each looking at a different issue. Groups discuss a case study and choose activities from pre-prepared cards. Groups feed back in plenary.<br>Handouts provided.<br>**90 mins** |
| | | **Session 2: Objective setting**<br>Participants understand how to set protection objectives, decide on outcomes to make people safer, and activities to deliver these outcomes | Small group discussion with trainer input focusing on response to specific threats identified in Module 1. Participants use a pre-set format to plan a programme.<br>Handouts provided.<br>**60 mins** |
| | | **Session 3: Indicators and monitoring**<br>Participants design indicators for monitoring progress and change in their programmes | Remaining in small groups, participants use pre-set formats to design indicators and monitoring processes for their programmes, followed by feedback in plenary.<br>Handouts provided.<br>**60 mins** |

# Example agendas

## Example three-day agenda

day 1

| Time | Session | Notes |
|---|---|---|
| 9.00 | Opening session | Introductions with inspirational quotes exercise, go through objectives and agenda. Explain and agree objectives and style of working. |
| 9.30 | What is protection? | Role play and presentation with visual aids |
| 10.00 | Tools for protecting civilians in conflict | Presentation and true or false exercise |
| 10.40 | Who protects? | Group exercise (optional true or false exercise) |
| 11.15 | **Break** | |
| 11.45 | Vulnerability analysis | Group exercise and discussion |
| 12.30 | **Lunch** | |
| 13.15 | Protection analysis | Discussion in plenary, group work, and feedback |
| 14.15 | Gathering and managing information | Group work |
| 15.00 | **Break** | |
| 15.15 | Gathering and managing information | Exercise and feedback in plenary |
| 16.00 | Agree/disagree | Exercise in plenary |
| 16.30 | Summary | |
| 17.00 | **Close** | |

day 2

| Time | Session | Notes |
|---|---|---|
| 9.00 | Opening session | Recap, go through agenda, questions |
| 9.30 | Risk analysis | Group work and feedback in plenary |
| 10.30 | Mainstreaming protection | Group work and feedback in plenary |
| 11.15 | **Break** | |
| 11.45 | Options for mainstreaming protection | Group work and feedback in plenary |
| 13.15 | **Lunch** | |
| 14.00 | What would you do if? | Exercise in plenary |
| 15.00 | True or false | Exercise in plenary |
| 16.00 | Summary | |
| 17.00 | **Close** | |

# day 3

| Time | Session | Notes |
|---|---|---|
| 9.00 | Opening session | Recap, go through agenda, questions |
| 9.30 | Options for protection programming | Group work and feedback in plenary |
| 11.00 | **Break** | |
| 11.30 | Objective setting | Presentation in plenary and group work |
| 12.30 | **Lunch** | |
| 13.15 | What would you do if? | Exercise in plenary |
| 13.45 | Indicators and monitoring | Group work and feedback in plenary |
| 15.00 | **Break** | |
| 15.30 | Agree/disagree | Exercise in plenary |
| 16.00 | Summary | |
| 16.30 | Evaluation and closing | |
| 17.00 | **Finish** | |

## Example one-day agenda

| Time | Session | Notes |
|---|---|---|
| 9.00 | Opening session | Introductions, go through objectives and agenda and style of working |
| 9.30 | What is protection? | Role play and presentation with visual aids |
| 10.00 | Tools for protecting civilians in conflict | Presentation and true or false exercise |
| 10.40 | Who protects? | Group exercise and true or false exercise |
| 11.15 | **Break** | |
| 11.45 | Protection analysis | Small group work and feedback in plenary |
| 12.30 | **Lunch** | |
| 13.15 | Risk analysis | Group exercise and discussion |
| 14.00 | Agree/disagree | Discussion in pairs and feedback in plenary |
| 15.00 | **Break** | |
| 15.30 | What would you do if? | Work in pairs and feedback in plenary |
| 16.15 | Summary | Questions and answers, summarise the day |
| 17.00 | **Close** | |

# Opening the workshop

Opening the workshop well is important so that:

■ the trainers know the level of knowledge and expectations of the group

■ the participants get to know each other and are clear about the structure and purpose of the training

This section includes some suggestions for:

■ testing the participants' level of knowledge

■ introducing people and setting the tone of the training

■ explaining the agenda

# Pre-training questionnaire

At the start of the training you may want to find out the level of knowledge of the participants. This will help you pitch the training at the right level. Using the questionnaire before and after the training (see 'Closing the workshop' p. 173) can help you to measure how much participants have learnt.

It is important that you explain to participants that this questionnaire is not an exam or a judgement on their performance and that their answers will not be shared with anyone else. It is simply to help the trainers understand what the group already knows, needs to learn, and (if using the questionnaire at the end of the training too) how effective the training has been in teaching it. It is also a way for participants to assess themselves and to judge how much they have learnt during the workshop.

You many decide to give certificates of participation at the end of the training (see 'Closing the workshop' p. 173). It is important that you explain that these certificates are not linked to the questionnaire – they are certificates of participation in the training, not of passing an exam or becoming 'accredited' to do protection work.

The trainer can use the questionnaire to see how participants have used their knowledge three or six months after the training. One way of doing this would be to repeat the questionnaire at the end of the training and ask the participants to fill in the final question on a postcard, which the facilitator collects and, where practical, sends back to them three or six months later. Where this is not practical the trainer could give postcards to the participants' line manager to give back to them at a later date. This is a way of checking not only that they have learnt something but also that they have also been able to use the knowledge.

## Trainer's notes

You will need:

1   1 copy of the questionnaire for each participant

2   an envelope or box for participants to place their completed questionnaires in

⧗ 15 mins

Name:

Position and organisation:

Date and location of training:

# PART 1: Circle the statement you agree with most

1. Protection means:

a) Guaranteeing the well-being of communities

b) Improving the safety of civilians

c) Ensuring that people are not attacked

d) Don't know

2. The kinds of threats we think about in protection are:

a) Violence, coercion, and deliberate deprivation

b) Poverty, suffering, and exclusion

c) Disease, hunger, and thirst

d) Don't know

3. A civilian is:

a) Someone who is not taking an active part in hostilities

b) Someone who does not have a weapon

c) Someone who is not wearing an army uniform

d) Don't know

4. If you want to find out about protection problems in a community you should always:

a) Ask police permission in advance

b) Get the names and details of individual cases

c) Talk confidentially with a number of different people

d) Don't know

5. A humanitarian organisation would never in any circumstances respond to a protection threat by:

a) Publicly criticising a government's behaviour

b) Visiting a threatened community more often

c) Calling for a community to be given weapons to defend themselves

d) Don't know

## PART 2: Circle the number to show how much you agree with the statement

1. I am confident that I know what protection means

NOT AT ALL    1      2      3      4      5      COMPLETELY

2. I have used a protection approach in my current work

NEVER      1      2      3      4      5      ALWAYS

3. 'Doing' protection is part of my job

NEVER      1      2      3      4      5      ALWAYS

4. I will be using a protection approach in my job in the next six months

NEVER      1      2      3      4      5      ALWAYS

[The following should be answered when completing the questionnaire at the end of the workshop]

If you answered statements 2–5 please give three examples of actions you will take to use a protection approach in your work in the next six months.

1._____

2._____

3._____

# Inspirational quotes

This exercise can be used as an informal way to help participants and trainers get to know a bit about each other at the start of a workshop. It creates a sense of shared values and a motivating atmosphere for the workshop.

## Trainer's notes

To deliver this session you will need:

1  to arrange the quote cards on a table or wall

2  to make a sign asking participants to chose a quote to read out when they introduce themselves

3  to provide blank post-it notes for those who wish to write out their own quote or proverb

## Resource materials for trainers

More inspirational quotes can be found on the Internet by searching for 'inspirational quotes' or 'motivating quotes'. Always aim to have a range of quotes from different individuals, cultures, and parts of the world.

⧗ 20–30 mins

## Session plan

As they arrive, ask participants to choose a quote to read out when they introduce themselves. Be sure to choose a quote yourself. Provide blank post-it notes or cards so that participants who already have a favourite quote or proverb can use it instead of the quotes provided.

Once you have opened the workshop and are ready for introductions, ask the participants to introduce themselves with their name, and organisation (if relevant), and to read out their quote, briefly explaining why they have chosen it.

Once all participants have introduced themselves you need to collect up the cards if you intend to use them again, or let participants take them home if you don't need them again.

# ✂ Inspirational quote cards

Every man of conviction must decide on the protest that best suits his convictions, but we must all protest.

Martin Luther King Jr

We don't need food, we need safety.

Iraqi girl in Safwan

Never doubt that a small group of thoughtful, committed citizens can change the world; indeed, it's the only thing that ever has.

Margaret Mead

He who allows oppression shares the crime.

Erasmus Darwin

A brave heart is a powerful weapon.

Anon

The only thing it takes for evil to triumph, is for good men to do nothing.

Edmund Burke

Peace is not something you wish for; it's something you make, something you do, something you are, something you give away.

Robert Fulghum

Peace is a daily, a weekly, a monthly process, gradually changing opinions, slowly eroding old barriers, quietly building new structures.

John F. Kennedy

First, do no harm.

Hippocrates

Peace is a journey of a
thousand miles and it must
be taken one step at a time.

Lyndon B. Johnson

Be patient and calm – for no
one can catch fish in anger.

Herbert Hoover

The time is always right
to do what is right.

Martin Luther King Jr

The future belongs to those
who believe in their dreams.

Eleanor Roosevelt

As we are liberated from
our own fear, our presence
automatically liberates
others.

Nelson Mandela

For to be free is not merely to
cast off one's chains, but to
live in a way that respects
and enhances the freedom
of others.

Nelson Mandela

There can be no keener
revelation of a society's
soul than the way in which
it treats its children.

Nelson Mandela

The future depends on what we do in the present.

Mahatma Gandhi

Peace cannot be kept by force; it can only be achieved through understanding.

Albert Einstein

If you want to make peace, you don't talk to your friends. You talk to your enemies.

Mother Theresa

Be the change you want to see in the world.

Mahatma Gandhi

Injustice anywhere is a threat to justice everywhere.

Martin Luther King Jr

The true measure of a man is how he treats someone who can do him absolutely no good.

Ann Landers

The more you sweat in peacetime, the less you bleed during war.

Chinese proverb

No tree has branches so foolish as to fight among themselves.

Ojibway Tribe

| Seeing nothing is as political an act as seeing something.<br><br><br>Arundhati Roy | Peace cannot exist without equality.<br><br><br>Edward Said |
|---|---|
| It is better to light a candle than curse the darkness.<br><br><br>Chinese proverb | An empty bag cannot stand.<br><br><br>Krio proverb |
| However long the night, the dawn will break.<br><br><br>African proverb | The only difference between stumbling blocks and stepping stones is the way you use them.<br><br>American proverb |

# Introducing the training

The opening session of the training is an opportunity for the participants to understand and agree the objectives of the training, the agenda, the roles of the trainers, and the style of working.

## Setting out the objectives

The main objectives of the training are for participants to be able to understand the basic principles of protection, know what it means in practice, and feel able to apply it in their work as appropriate to their role. The specific objectives of each module are set out on the opening pages of each module and the trainer can draw these out depending on which modules they are using. The trainer should take a few minutes at the start of the training to explain the objectives to the participants and to see if they have additional objectives or expectations to add.

## Setting the agenda

The agenda you choose for the training will depend on the level of knowledge and experience of the participants, the time available, and the objectives of the training. Whatever agenda you choose, you should present it clearly to the participants at the start of the training and keep referring back to it in the course of the training so that participants are reminded about what they have covered and what remains to cover. Different trainers will have their own preferred ways of presenting the agenda.

Overleaf is one method that has been popular in protection trainings around the world. It shows the training agenda as a journey. The trainer draws it on large paper in bright colours before the training and explains it at the start of the workshop. The agenda stays on the wall throughout the training and the trainer uses it in the summaries at the beginning and end of each day.

## Explaining roles

This training requires at least two trainers. Usually they will have different backgrounds, experience, knowledge, and skills and will take different roles throughout the training. One may be responsible for providing explanations in plenary, while the other might facilitate the small group work. One might be an international protection specialist while the other will be a national or local protection practitioner, able to provide appropriate practical examples, context, and language. Explain these different roles to the group so that they know what to expect from each trainer.

## Style of working

At this point the trainer and participants can suggest and agree on some basic ground rules to ensure that the training runs smoothly, time and opinions are respected, and disruptions are kept to a minimum. This training is highly interactive and asks participants to bring in and apply learning to their own experiences throughout the modules. The trainer should explain this from the outset. The trainer can also use this opportunity to explain the concept of the 'talking wall', a wall on which key visual aids and explanations will be stuck during the course of training for participants to keep referring back to.

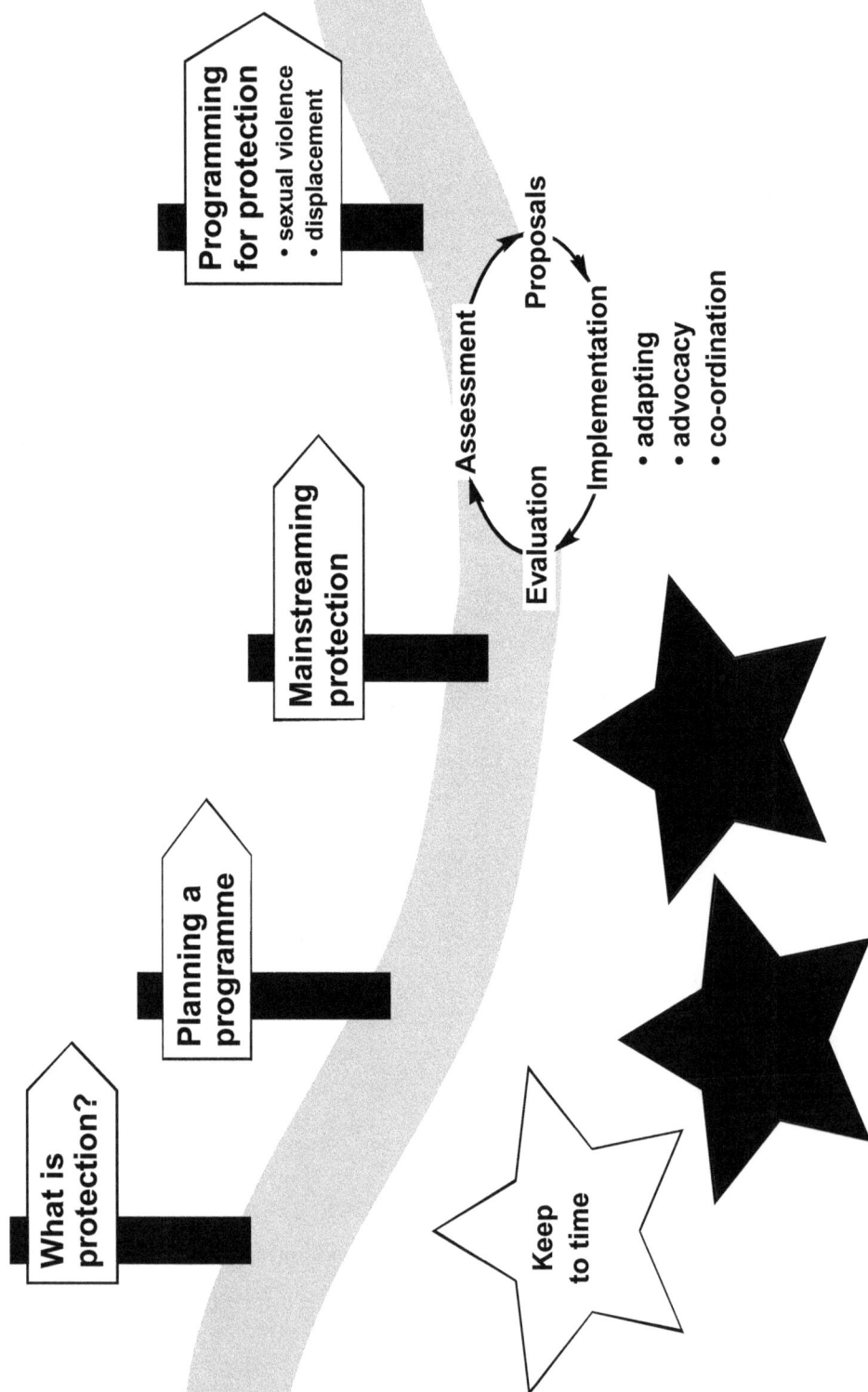

# Agenda for protection training

**What is protection?**

**Planning a programme**

**Mainstreaming protection**

**Programming for protection**
- sexual violence
- displacement

Assessment — Proposals

Evaluation — Implementation

- adapting
- advocacy
- co-ordination

**Keep to time**

# Module 1
## What is protection?

This module
- looks at what 'protection' means
- introduces some of the concepts and tools used in protection work

Session 1 looks at what is included in protection, and at different types of action that humanitarian organisations can undertake to protect people.

Session 2 looks at how the law and other standards can be used in protection.

Session 3 identifies who are the main people and organisations involved in protection work, and their respective roles and responsibilities. It looks at the central role that communities take in protecting themselves and also emphasises why it is important for protection actors to co-ordinate and co-operate.

Session 4 helps participants understand more about vulnerability, introducing the idea of power analysis.

# Session 1: What is protection?

This session begins with a simple role play followed by small group discussion and a presentation with visual aids.

## Trainer's notes

To deliver this session you will need:

1  to brief your actors for the role play (ask two participants to act it out using the briefing provided)

2  the visual aids (provided on pages 34–7 and in colour at the back of the book)

3  a flip chart and pens

4  copies of Handout 1 – What is protection? (p. 33)

You will also need blu-tak to stick the visual aids to the wall.

## Resource materials for trainers

▨ Handout 1: What is protection?

▨ H. Slim and A. Bonwick (2005) *Protection: An ALNAP Guide for Humanitarian Agencies*, London: Overseas Development Institute, Section 1.

⧖ 30 mins

# Session plan

**Begin with the role play. After the role play, use the visual aids provided, and tell participants that:**

- Protection is about improving the safety of people, like the woman collecting water. There are two sides to protection – reducing the negative consequences of our actions, and proactively helping people stay safer. This workshop focuses on the latter.
- Humanitarian organisations help people stay safer by reducing risk.
- Risk exists when there is a threat, such as the threat of violence or exploitation, and people are vulnerable because, for example, they are female, are from a certain ethnic group, or because they have to leave their village to get water. The longer the time people are exposed to a threat the greater the risk – for example collecting water four times a day is riskier than collecting it twice a week.

## There are three types of threats:

- **Violence** – deliberate killing, wounding, sexual violence, rape, torture, and the threat of any of these
- **Coercion** (forcing someone to do something against their will) – forced prostitution, sexual slavery, sexual exploitation, forced or compulsory labour, forced displacement or return, restriction of movement, prevention of return, forced recruitment, being forced to commit acts of violence against others
- **Deliberate deprivation** – destruction of homes, wells, and clinics; preventing access to land or markets; preventing delivery of relief supplies; deliberate discrimination in getting jobs, education, land, or services; illegal 'taxes' or tolls

## NGOs try to improve the safety of people by doing one or more of:

- reducing the level of threat
- reducing the level of vulnerability
- reducing the amount of time exposed to the risk

Ask the participants to think about the role play and suggest ways that an organisation might try to protect the woman collecting water. Write up the suggestions on a flip chart. Explain that there are various actions an organisation might take to reduce the threat or reduce vulnerability:

## To reduce the threat

- **Advocacy:** convincing those with power to protect people or getting others to put pressure on them to protect people
- **Capacity-building:** supporting the authorities to protect civilians
- **Presence:** using physical presence to deter attacks on civilians (can also reduce time exposed to the risk)

## To reduce vulnerability

- **Assistance:** directly providing services or goods so that people can avoid threats (can also reduce time exposed to the risk)
- **Voice:** helping people to negotiate their own safety
- **Information:** providing impartial information to help people make informed decisions about their safety

You can write these six modes of action on a flip chart to form part of the talking wall, and keep referring back to it throughout the training. Pick an example of each mode of action from the participants' suggestions on the flip chart. At this point, and throughout the training, encourage participants to think about appropriateness and risks involved in each type of action.

In the next session we will look at how the law and other standards can be used in protection.

# Role play briefing

## Characters

- A woman collecting water (props: bucket or water container)
- A man with a gun (props: roll of paper or umbrella)

It is important that

1  the woman mentions how many times a day she has to collect water, as this illustrates how often people are exposed to risk.

2  the man (1) refuses to let her pass, (2) threatens to beat her, and (3) asks her to be his 'friend', as these three acts illustrate the main types of threat.

## Suggested dialogue

**Woman:**  Oh no, I have to collect water again, this is the fourth time today, but I must go because my family is desperate and the well in our village is very old and the water not good for drinking.

**She meets the armed man.**

**Man:**  Stop, you can't come past here. This is our well.

**Woman:**  But that is the well our village uses. It is the only place where we can get water. You must let me pass.

**Man:**  Don't tell me what I must do. This is our well now, stop complaining or I will beat you.

**Woman:**  Please let me pass, my daughter is sick and I need water for her. Please.

**Man:**  Well, let's see, maybe if you come into my hut and you are very nice to me, I will let you have a little water for your daughter.

# What is protection?

Protection is about improving the safety of civilians.

Where there is a **threat** and people are **vulnerable** they are at **risk**.

The more time people face the **threat**, the higher the **risk**.

## Threat + Vulnerability x Time = RISK

> **Example:**
> A woman goes out of her village to collect water. A man blocks her way and threatens her with violence.
> The actions of the man are the threat.
> The woman may be vulnerable because she is a woman, or from a certain ethnic group, and also because she has no water source in her village.
> The more times she has to go and collect water the greater the risk to her.

Threats include:

- **Violence** – deliberate killing, wounding, torture; cruel, inhuman, and degrading treatment; sexual violence including rape; the fear of any of these
- **Coercion** – (forcing someone to do something against their will) – forced prostitution, sexual slavery, sexual exploitation, forced or compulsory labour, forced displacement or return, restriction of movement, prevention of return, forced recruitment, being forced to commit acts of violence against others
- **Deliberate deprivation** – destruction of homes, wells and clinics; preventing access to land or markets; preventing delivery of relief supplies; deliberate discrimination in getting jobs, education, land, or services; illegal 'taxes' or tolls.

## Reducing risk

Non-government organisations (NGOs) try to reduce risk by reducing the threat, reducing the vulnerability, and reducing the time people face the threat. NGOs work in co-ordination with others to do some or all of these:

**To reduce the threat**
- **Advocacy:** convincing those with power to protect people or getting others to put pressure on them to protect people
- **Capacity-building:** supporting the authorities to protect civilians
- **Presence:** using physical presence to deter attacks on civilians

**To reduce vulnerability**
- **Assistance:** directly providing services or goods so that people can avoid threats
- **Voice:** helping people to negotiate their own safety
- **Information:** providing impartial information to help people make informed decisions about their safety

## Templates for visual aids

These visual aids are for use in Session 1. The visual aids should be placed on a wall where they will remain as a reference point for the duration of the workshop. Copies can be printed from the CD, or the text can be written on flip charts. There are also larger colour versions at the back of the book.

Improving the safety of civilians

Jane Beesley/Oxfam

**Protection encompasses all activities aimed at obtaining full respect for the rights of the individual in accordance with the letter and the spirit of the relevant bodies of law**

ICRC Workshop, 1999

Threat +

Vulnerability

x Time

= RISK

# Deprivation

**Stopping people from accessing the goods and services they need to survive**

Examples: the destruction of homes, crops, wells, clinics, and schools; preventing access to land and markets; preventing the delivery of relief supplies; deliberate discrimination in access to property, land, jobs, and services; illegal 'taxation' or tolls

# Coercion

**Forcing someone to do something against their will**

Examples: forced prostitution, sexual slavery, sexual exploitation, forced or compulsory labour, forced displacement or return, restriction of movement, prevention of return, forced recruitment, being forced to commit acts of violence against others

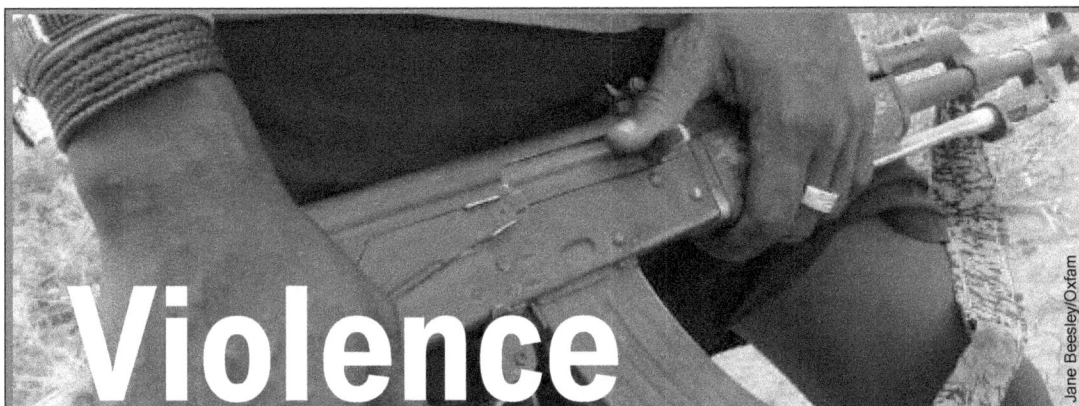

# Violence

## or the fear of violence

**Examples: deliberate killing, wounding, and torture; cruel, inhuman, and degrading treatment; sexual violence including rape; the threat or fear of any of the above**

Jane Beesley/Oxfam

# Session 2: Tools for protecting civilians in conflict

This session is an interactive presentation of the key standards and tools for humanitarian protection.

## Trainer's notes

To deliver this session you will need:

**1** copies of Handout 2: International standards for civilian protection (p. 44)

**2** blu-tak to stick up the poster (a large version of Handout 2 – available in colour at the back of the book).

It is also advisable to provide basic information on the standards and principles listed here for participants to refer to in future. Shortened or simplified versions of many of these documents are available on the Internet, including on the sites of ICRC, OCHCR, UNHCR, and other international organisations.

If you are facilitating this session you should have a basic understanding of these standards and principles to help you explain them and answer questions from participants. In many conflict contexts, participants are very well informed and have lots of opinions about the gap between standards and reality. The trainer will need to be aware of sensitive and controversial areas and be prepared to handle them.

Below is a very brief summary of the standards and principles for trainers to refer to when they are presenting the poster and guiding discussion. It is not intended as a script for the trainer as it is quite long.

### International and regional declarations and conventions on human rights

**The Universal Declaration of Human Rights (1948)** is the foundation of human rights. It is a declaration which was followed and complemented by two conventions:

▪ **The International Covenant on Civil and Political Rights (1966)** protects the civil rights and liberties of individuals from violations by the state authorities. It includes the rights to freedom of movement and the right to equality before the law. States have the duty to ensure these rights are fulfilled and not to violate them. In times of war or emergency, states can choose to derogate from (to suspend) certain of these rights. However, there are some that are so important that even in times of emergency, governments have to uphold them.
These are:
  - the right to life
  - the right not to be tortured or treated in an inhuman or degrading manner
  - the right to freedom of thought and religion
  - the prohibition of slavery

- **The International Covenant on Economic, Social and Cultural rights (1966)** requires that states take measures to ensure the well-being of each individual. It includes the rights to education and to health.

  There are also several regional conventions which have been agreed to reinforce these international rights and to apply them to specific regional contexts. These include the **African Charter on Human Rights (1986), The European Convention on Human Rights (1950), and the American Convention on Human Rights (1978).**

There are several thematic conventions which deal with specific aspects of human rights. These include:

- The Convention against **Torture** and all forms of Inhuman and Degrading Treatment (1987)
- The Convention on the Elimination of all forms of **Discrimination against Women** (1981)
- The Convention on the **Rights of the Child** (1990), which also has an optional protocol on children in armed conflicts (2000), which prohibits the recruitment or participation in conflict of children under 18 years old
- The Convention on the Prevention and Punishment of **Genocide** (1951)

There are committees to which states must periodically report their progress towards ensuring rights are respected and fulfilled. There are Committees on the Conventions on Civil and Political Rights, Rights of the Child, and the Conventions against Discrimination against Women and against Torture. States have to submit reports to these Committees approximately every four years. States can also submit complaints against another state to these committees, and individuals and NGOs can submit complaints to the African Commission on Human Rights. There is a big gap around sanctions against states which violate human rights. Committees can condemn states but there are no mechanisms to punish them.

## International and regional instruments for protecting refugees and internally displaced people

- **The Refugee Convention** was agreed in 1951. According to the Convention, a refugee is someone who has fled their country and who has crossed an international border, due to persecution. The UN High Commissioner for Refugees (UNHCR) is responsible for promoting that states sign and implement the Convention. The Convention sets out the rights of refugees and the responsibilities of host countries. Among these are the right to seek asylum in another country and to have one's case examined by a competent authority. For those who are accepted as refugees, they have the right to receive the same treatment as the host community – with access to education and freedom of movement, for example. The most important article in the Convention concerns the right to 'non-refoulement' – this means that a country cannot send someone back to a place where they would risk harm or death. Even in war when neighbouring countries receive large numbers of refugees, it is forbidden to close the borders to new refugees.

- **The OAU Convention (1974)** applies the 1951 Convention to the African context. It expands the definition of refugees to include people who have fled their homes due to war and conflict and not just due to individual persecution.

■ Internally displaced people (IDPs) are people who have fled their homes for the same reasons as refugees but who have not crossed an international border. Because IDPs are not protected by refugee law, the **UN** set out the **Guiding Principles on Internal Displacement**. These principles are drawn from human-rights and refugee law principles. They are not legally binding but they do have a considerable authority and some countries have taken the step of integrating them into their national law. They set out standards for the assistance and protection of IDPs during displacement, return, and reintegration, according to principles of dignity, voluntariness, and access to impartial information.

## International humanitarian law

■ **The Geneva Conventions** are the basis of **international humanitarian law (IHL)** and apply during war. There are four Geneva Conventions and two additional protocols. The 4th Geneva Convention concerns the protection of civilians during war. A civilian is someone who does not take part in the hostilities – and combatants are prohibited from targeting them. The 2nd protocol concerns rights and responsibilities during internal (as opposed to international) armed conflicts.

■ **War crimes** are grave violations of international humanitarian law, including rape, recruitment of children, violence targeted against civilians, the taking of hostages, and blocking or targeting aid workers or supplies.

■ Individuals who commit war crimes can be prosecuted by the **International Criminal Court (ICC)** which was created in 2002 by the Rome Statute. The Court has jurisdiction over war crimes, genocides, and crimes against humanity that have been committed since 2002. Crimes against humanity can include those committed outside conflict and are violations of human rights which are grave, widespread, and systematic. The court is complementary to national courts – it can prosecute cases that states are unable or unwilling to prosecute in their national courts. Proceedings against an individual can be initiated by a state, or by the UN Security Council, or by the Chief Prosecutor of the ICC.

## The UN Charter and other UN documents and resolutions

■ **The UN Charter** is the treaty which established the UN in 1945. It has the highest status of all the international conventions. The charter has two objectives:
- Maintaining peace and security through a system of collective security
- Promotion of the social and economic advancement of all people, through the respect of human rights and co-operation between states.

The charter bans the use of armed force by states – except in cases of self-defence. Even in cases of self-defence states must seek peaceful solutions. On the principle of collective defence, the charter authorises the international community (the UN Security Council) to take steps to maintain international peace and security. Under Chapter 6 of the UN Charter, peaceful measures to resolve a dispute can be taken, but if these fail, under Chapter 7 the UN Security Council can agree coercive measures. The first coercive measure is to impose collective sanctions (diplomatic, military, or economic) against a state which poses a threat to international peace and security. The second is the deployment of peacekeeping or peace-enforcement forces.

- **Mandates** of peacekeeping or peace-enforcement missions explain their role in a given country. Very often these include a mandate to protect civilians, though the scope of this mandate and the means permitted to enforce it vary considerably between missions.

- The **Secretary General's Bulletin on the Protection against Sexual Abuse and Exploitation** was written in 2003 to respond to accusations in West Africa that humanitarian workers had sexually exploited women and girls. The bulletin prohibits all those working with or for the UN (NGO partners included) from having sexual relations with beneficiaries, paying for sex with goods or money, and having sexual relations with anybody under 18.

- In 2005 the UN Summit agreed a declaration on the **Responsibility to Protect**. Governments reaffirmed that the duty to protect populations from genocide, war crimes, ethnic cleansing, and crimes against humanity rests first and primarily with each state and that this duty involves taking measures to prevent these crimes. In addition, governments at the Summit reaffirmed that the international community, by way of the UN, equally has a responsibilty to use peaceful means to protect civilian populations against genocide, war crimes, crimes against humanity, and ethnic cleansing. They expressed their will to react multilaterally against these threats when states do not protect their own populations. This international intervention includes collective action by way of the UN Security Council. The international community is equally engaged in helping states to build their own capacity to avoid conflicts and protect their populations.

## National law

All states have the duty to ensure human rights are fulfilled and are not violated. These rights must be reflected in national law and be put into practice by means of the civil authorities, the justice sector, the police, and the armed forces.

## NGO standards for self-regulation

- **The Red Cross Code of Conduct** was created in 1994 by the Red Cross and Red Crescent movement and several humanitarian NGOs. This optional code of conduct sets out ten principles, including:
  - **Humanity** – priorities for delivering humanitarian assistance are based on needs
  - **Impartiality** – humanitarian assistance is delivered without discrimination on grounds of the religion, ethnicity, or gender of beneficiaries
  - **Independence** – humanitarian agencies ensure that they retain operational control and direction of humanitarian activities. They formulate and implement their own policies independently of government policies or actions.

- The **Sphere Standards** and the **Humanitarian Charter** were created by humanitarian organisations like the Red Cross and several NGOs in 1997 to reaffirm and define minimum standards to maintain the life and dignity of civilians in situations of emergency or conflict. The Sphere Minimum Standards outline the quality requirements for humanitarian assistance. The Minimum Standards are a practical expression of the principles and rights embodied in the Sphere Humanitarian Charter. The Humanitarian Charter draws on international human-rights and humanitarian law to set out the most basic requirements for sustaining the lives and dignity of those affected by calamity or conflict.

In particular the Sphere Humanitarian Charter champions the key legal principles of:
- Right to life with dignity (international human-rights law)
- Distinction of civilians from combatants (international humanitarian law)
- Non-refoulement and ban on forcible or coerced displacement (refugee law, the Internally Displaced Persons Guidelines, and international human-rights law)

◾ Most humanitarian organisations will have their own **codes of conduct**. For example, all Oxfam staff sign the **Oxfam Code of Conduct**. In this Code, staff agree among other things to:
- Treat all individuals with equality, respect, and dignity
- Respect local laws unless the Code sets a higher standard than local law; then the Code will apply
- Not to have any transactional sexual relations with beneficiaries. According to the Code, this includes all exchange of money, goods, services, and favours for sex with any other person
- Not to have transactional sexual relations with anybody under the age of 18
- Not to have sexual relations that bring Oxfam GB into disrepute

## Resource materials for trainers

◾ Handout 2: International standards for civilian protection

◾ F. Bouchet-Saulnier (2006) *The Practical Guide to Humanitarian Law*, Lanham MD: Rowan & Littlefield Publishers.

◾ IASC Task Force on Humanitarian Action and Human Rights (2004) 'Frequently Asked Questions on International Humanitarian, Human Rights and Refugee Law in the context of Armed Conflict' www.icrc.org/web/Eng/siteeng0.nsf/htmlall/637K8G/$file/FAQs.pdf

⧖ 40 mins

## Session plan

Protection is closely associated with the law. This session examines how rights and other standards can be used to improve the protection of civilians.

Begin with a presentation of the law and other standards, using a large version of Handout 2.
This can be done in an interactive manner by asking questions such as:

- Does the woman in the role play from session 1 have a right to access water? Why?
- Does the woman have a right not to be attacked by the man? Why?
- Does the woman have a right to collect water without having to 'be nice' to the armed man? Why?
- Who says she has these rights?
- Who or what determines what the armed man can or cannot do?
- What rules govern the actions NGOs take to protect the woman?

Ask the group to think back to the role play and discuss how these standards can be used to protect the woman:

- by defining acceptable behaviour
- by determining who is responsible for protecting civilians
- as a tool to persuade those with power to protect civilians

The trainer should emphasise that the law and other standards do not themselves protect people.
It is respect for the law and other standards that protects. However, enforcement of the law is often difficult, and people may be more willing to comply with the law if they can be shown that it is in their political interest to do so.

☞ **handout 2**

# International standards for civilian protection

## Refugee Law

The 1951 **Refugee Convention** sets the standard for the protection of refugees.

The **Cartagena Declaration on Refugees** and the **OAU Refugee Convention** provide enhanced protection to refugees in Latin America and Africa respectively.

## International Humanitarian Law

The **Geneva Conventions** and Additional Protocols protect people not (or no longer) taking an active part in the hostilities. Hague Law is made up of treaties that restrict the means and methods of warfare to prevent unnecessary losses or suffering.

The **International Criminal Court** has jurisdiction over genocide, crimes against humanity, and war crimes.

## The International Bill of Human Rights

- Universal Declaration of Human Rights
- International Covenant of Civil and Political Rights
- International Covenant of Economic, Social and Cultural Rights

**Regional human-rights laws include:** the European Convention on Human Rights, the African Charter on Human and People's Rights, and the Inter-American Convention on Human Rights.

The **United Nations** Charter bans the use of force between states except in self-defence or when authorised by the Security Council.

Every **state** is responsible for protecting people on its territory. International standards should be reflected in national law and implemented through the public administration, judiciary, police, and armed forces.

The **Genocide Convention** demands that states prevent and punish the crime of genocide.

The **Torture Convention** bans the use of torture in all circumstances.

The **Convention on the Rights of the Child** provides special protection and rights for children. The **Optional Protocol on Children in Armed Conflict** adds standards for protecting children in armed conflict.

The **Convention on the Elimination of all forms of Discrimination Against Women** established minimum standards to end discrimination against women.

The **UN Guiding Principles on Internal Displacement** provide non-binding but authoritative guidance on assistance to and protection of internally displaced people during displacement, return, and resettlement.

The **2005 UN World Summit** reasserted state responsibility for protecting its population, and also responsibility of the international community to help protect people from genocide, war crimes, ethnic cleansing, and crimes against humanity.

**UN Peacekeeping or Peace-Enforcement Missions** are often mandated to protect civilians, and to promote human rights.

The **Secretary General's Bulletin** on protection from sexual abuse and exploitation governs the ethical conduct of UN personnel.

NGOs have also developed voluntary standards for self-regulation:

- **The Humanitarian Charter** asserts the principles of humanity, impartiality, and independence.
- The **Red Cross Code of Conduct** sets out principles for their behaviour and actions.
- The **Sphere Minimum Standards** outline quality standards for humanitarian aid.

# Session 3: Who protects?

This session uses small-group work to identify how people stay safe in armed conflict, who provides protection, what people do themselves in order to stay safe, and what role humanitarian organisations play in protection.

## Trainer's notes

To deliver this session you will need:

1 a set of 'Who protects?' cards. These cards consist of large cards representing various actors and small cards detailing different roles relating to those actors. They are provided on pages 47–54, and are available in colour at the back of the book. Divide the participants into two groups. Give one group cards relating to the UN Security Council, Other states, The state, and Non-state armed actors (black cards). Give the other group cards relating to ICRC, UNHCR, NGOs, and Communities (white cards). If you have a very large group use two sets of cards and form four groups.

2 copies of Handout 3: Who protects? (p. 55)

You will also need blu-tak and a piece of flip chart paper to stick the cards up.

## Resource materials for trainers

Handout 3: Who protects?

H. Slim and A. Bonwick (2005) *Protection: An ALNAP Guide for Humanitarian Agencies*, London: Overseas Development Institute, pp. 30–9.

It may be useful to have a 'resource' person in each group to assist the groups in completing the task.

## ⧗ 30–60 mins

## Session plan
### (Group work 15–30 mins; feedback and discussion 15–30 mins)

### Group work:

1 Explain that there is a range of people involved in or responsible for protecting civilians. In this exercise we are going to identify who are the main 'protection actors', what responsibilities they have, and what actions they may take when faced with a threat.

   Give each group a set of cards and ask them to stick the four large cards across the top of a flip chart. They must then stick the smaller cards under the relevant protection actor.

2 Monitor each group to ensure that they understand the task and are able to carry it out.

3 When the groups have finished, ask them to look at the other group's flip charts and comment on them. Add in your own comments or use questions (e.g. 'Who is legally responsible?'; 'Who do you think does the most to protect civilians?' etc.) as necessary to ensure that the learning points (below) are covered. Put the two flip charts together to show the full range of protection actors. Write in any extra actors or roles that the group come up with during discussion.

4 Give out Handout 3 to each participant.

### Key points:

▨ Most protection is 'self-protection' by individuals, families, and communities.

▨ Overall legal responsibility for the protection of civilians lies with states as signatories to international humanitarian law (IHL), human-rights law, and refugee law. In internal conflicts non-state actors must abide by IHL.

▨ The law in itself is not protective; compliance with and respect for the law is what protects.

▨ Some agencies have special mandates to protect: ICRC, UNHCR (and other UN agencies).

▨ Peacekeeping forces may have a mandate to protect civilians.

▨ Humanitarian organisations also contribute to humanitarian protection through supporting and strengthening 'self-protection' and holding governments and other authorities to account. Use the cards to illustrate the types of actions humanitarian organisations might take and to show the relationship between the different actors.

**The second set of true or false questions in the Core exercises on p.144 has been designed to reinforce learning from this session and can be used immediately following this session or later in the workshop.**

# 'Who protects?' large cards

**United Nations Security Council**

**Non-state armed actors**

## Communities, families, individuals, tribes

## Non-government organisations

**United Nations High Commissioner for Refugees**

**Other states including
European Union, African Union, ECOWAS**

## International Committee of the Red Cross

# THE STATE

| Police force | Military | Judiciary |

# 'Who protects?' cards (white)

| | |
|---|---|
| Avoid threats e.g. by fleeing, hiding, not travelling, etc. | Submit to threats e.g. paying illegal 'taxes' |
| Confront threats e.g. protesting, joining militias, arming themselves | Negotiate their own safety with armed actors |
| Solidarity – practical help and support to friends and family | Advocate on behalf of civilians |
| Encourage states to sign up to international humanitarian law | Legally mandated to promote and disseminate international humanitarian law |
| Mandated to co-ordinate action to protect refugees and resolve refugee problems worldwide | Encourage states to promote protection of human rights and the peaceful resolution of disputes |

Build the capacity of
state organs (e.g. army)
to implement international
humanitarian law

Endeavour to ensure the
protection of and assistance
to military and civilian
victims of armed conflict

Maintain a presence to
discourage threats to civilians

Provide impartial assistance

Build the capacity
of communities to
protect themselves

Provide information to
people about their options

Give voice to civilians

UN global lead on protection

Assist the reintegration
of returning refugees in
their country of origin

Major responsibility for
the protection of internally
displaced people

# 'Who protects?' cards (black)

| | |
|---|---|
| Apprehend and punish perpetrators | Legal responsibility for upholding international humanitarian law, human-rights law, and refugee law |
| Commanders and members of forces have a personal responsibility for violations of the law | Even though not a signatory, obliged to uphold international humanitarian law |
| Maintain international peace and security | Support states e.g. approving peacekeeping missions |
| Coercive measures e.g. peace enforcement and sanctions | Put pressure on member states to uphold their legal obligations |
| Defence of the nation | Primary responsibility for the protection of civilians on its territory |

| | |
|---|---|
| Protection of civilians in the area in which they operate | Put pressure on fellow (member) states to uphold their legal obligations |
| Provide troops for peacekeeping or carrying out own peacekeeping operations | Provide refuge for those fleeing persecution and conflict |
| Maintain the rule of law | Provide humanitarian aid to its citizens |
| Provide a safe environment for their people to go about their lawful business | Promote and respect human rights |

# Who protects?

☞ handout 3

| Communities, families, individuals, tribes | The state police force, military, judiciary | United Nations High Commissioner for Refugees | International Committee of the Red Cross | Non-government organisations | United Nations Security Council | Non-state armed actors | Other states, European Union, African Union, ECOWAS |
|---|---|---|---|---|---|---|---|
| Avoid threats e.g. by fleeing, hiding, not travelling, etc. | Provide a safe environment for their people to go about their lawful business | Mandated to co-ordinate action to protect refugees and resolve refugee problems worldwide | Encourage states to sign up to international humanitarian law | Maintain a presence to discourage threats to civilians | Maintain international peace and security | Commanders and members of forces have a personal responsibility for violations of the law | Put pressure on fellow (member) states to uphold their legal obligations |
| Submit to threats e.g. paying illegal 'taxes' | Apprehend and punish perpetrators | Major responsibility for the protection of internally displaced people | Legally mandated to promote and disseminate international humanitarian law | Build the capacity of communities to protect themselves | Support states e.g. approving peacekeeping missions | Protection of civilians in the area in which they operate | Provide refuge for those fleeing persecution and conflict |
| Confront threats e.g. protesting, joining militias, arming themselves | Maintain the rule of law | UN global lead on protection | Build the capacity of state organs (e.g. army) to implement international humanitarian law | Advocate on behalf of civilians | Coercive measures e.g. peace enforcement and sanctions | Even though not a signatory, obliged to uphold international humanitarian law | Provide troops for peacekeeping or carrying out own peacekeeping operations |
| Negotiate their own safety with armed actors | Provide humanitarian aid to its citizens | Assist the reintegration of returning refugees in their country of origin | Endeavour to ensure the protection of and assistance to military and civilian victims of armed conflict | Promote and respect human rights | Put pressure on member states to uphold their legal obligations | | |
| Solidarity – practical help and support to friends and family | Primary responsibility for the protection of civilians on its territory | Encourage states to promote protection of human rights and the peaceful resolution of disputes | | Provide information to people about their options | | | |
| | Defence of the nation | | | Provide impartial assistance | | | |
| | Legal responsibility for upholding international humanitarian law, human-rights law, and refugee law | | | Give voice to civilians | | | |

# Session 4: Vulnerability analysis

This session starts with a group exercise involving role play and moving around the room, followed by a facilitator-led discussion.

## Trainer's notes

To deliver this session you will need:

1 to prepare the 'Play the role of...' cards – ensuring a diversity of roles within the group (provided on pp. 58–64 and available in colour at the back of the book)

2 to photocopy the series of statements that you will read out to participants during the role play (p. 65)

The suggested roles are for a mixed nationality group. You can create your own roles that are more specific to your context and the country or region you are working in.

Make sure that the roles highlight the factors that make people vulnerable (e.g. gender, age, disability, ethnicity, location, social status, and so on). You can also adapt the list of statements to illustrate the power relations between different roles.

## Resource materials for trainers

- H. Slim and A. Bonwick (2005) *Protection: An ALNAP Guide for Humanitarian Agencies*, London: Overseas Development Institute, pp.51–5.

⧖ 45 mins

# Session plan

In this session participants will play the role described on their role card – they must keep their role a secret from other participants. Give out the 'Play the role of…' cards. Ask participants to spend a few minutes imagining the daily life of the person whose role they are playing and to think of a name for their character. Make sure that they understand that they will be answering questions in that role.

Some roles will not be culturally appropriate, or may be difficult for participants to play (e.g. a warlord, a prostitute, a woman who has been raped). Sort the cards in advance and remove any that are not appropriate. You may need to make up new cards to ensure there are enough for all participants. Alternatively you can ask other trainers to take part in the exercise and play these roles. Tell participants that they can change their card if they are very uncomfortable with the role they have been given, and keep some 'neutral' roles aside for this purpose.

Ask the participants to line up at one end of the room. Read out the list of statements on p. 64 and ask participants whose role can agree with the statement to take a step forward. Eventually participants will be spaced out across the room according to how many steps they have taken forward. Encourage those participants who are unsure how to answer to ask you for advice. Once you have read out all the statements, the participants stay where they are and introduce their role to the group. Ask the other participants if they think that role is correctly placed in relation to others (e.g. should the police officer be further forward than the orphaned girl?). In some cases this can create a lot of discussion, so allow plenty of time as the discussion is a very important part of the exercise. Ask the participants to identify what factors make some people more vulnerable to protection threats (some suggestions are included on p. 65).

# Vulnerability is about power

Choose two roles, and ask participants to say who is the more powerful of the two and why. Repeat this several times. This can be used to show how exploitation and abuse is based on power dynamics.

Explain to the group that a protection response aims to reduce vulnerability and to reduce both the level of threat and the time that people are exposed to the threat. Ideally it would do all three, but in reality it may be able to do only one or two. In the next session we will identify the threats.

# Table-top version

This game can be played on a table-top if you have limited space or participants have mobility limitations. Draw a grid on flip chart paper. Participants can think of a name for the role they are playing and write it on a small post-it note. If they can agree with a statement they move their post-it note forward on the grid.

## 'Play the role of' cards

Play the role of ... a 10-year-old girl. **You live in an orphanage in Kinshasa and rarely have enough to eat. Sometimes you are beaten.**

Play the role of ... an asylum seeker **in the UK. You have to work illegally as a labourer to repay the traffickers who brought you to the UK.**

Play the role of ... a humanitarian **worker on a national contract with a UN agency. You have a good salary and can meet your family's basic needs.**

Play the role of ... a community leader **from a dominant ethnic group with a rich and influential family.**

Play the role of ... a farmer. Your fields are in a dangerous area and you risk being wounded or killed if you go there. If you don't work the fields your family will go hungry.

Play the role of ... a married man with four children running a food centre in a small town. You lost an arm in the war, but have a good income.

Play the role of ... a 14-year-old girl head of household with two younger sisters to support. You struggle to survive and make a small income in any way you can, including prostitution. You are regularly beaten by your clients and recently you have become very ill.

Play the role of ... a disabled man from a minority ethnic group living in a village. You have no money or support and have to beg for food every day. You are often beaten by people in your village.

Play the role of... a policeman in the new Iraqi police force, married with two children. You receive a small salary, and with help from your family you are putting both your children through school.

Play the role of... a cleaner at a UN peacekeepers' base. You receive a small salary plus the opportunity to get some extra things or money from the soldiers if you become their 'special friend'.

Play the role of...a worker for a national human-rights NGO. You do a dangerous job and work very long hours for a small salary.

Play the role of... a grandmother living in a displacement camp. You have lost your family and don't want to return to your village because there is nothing there for you. You don't have enough food to eat.

Play the role of ... a woman displaced from your village and hiding in a forest. You forage for food and water but often go hungry and are very scared.

Play the role of ... a 16-year-old boy working in a diamond mine. You are given some food but no money. You have no family.

Play the role of ... a local government officer. You have influence and can afford everything you need.

Play the role of ... the President of Uganda.

Play the role of... the head of a UN agency in Darfur.

Play the role of... a woman returning to your village from a displacement camp. You have a return package from UNHCR but your land and house have been taken by someone else.

Play the role of... a man returning to Liberia after living in America for ten years. You have a Ph.D from Harvard University and money to invest in new business opportunities.

Play the role of... a businessman in Khartoum. You own several compounds rented to NGOs, but you have a big family to support.

Play the role of ... a prostitute in Goma. You earn $15 a night. You get sick often, and you are beaten and raped on a regular basis. You are pregnant.

Play the role of ... a former warlord living in a rubber plantation. You make a lot of money from the rubber and have influential connections.

Play the role of ... an internally displaced person returning to your town where you are in an ethnic minority. You can't get your land back so can't grow food for yourself or to sell.

Play the role of ... a refugee from Colombia living on the streets in Quito. You have no money, family, or friends.

Play the role of ... an international aid worker. You receive a good salary and live in a compound with a swimming pool.

Play the role of ... a woman living in a village with no water supply. Recently some armed men have come into the area and they won't let you get water from the nearest well.

Play the role of ... an 18-year-old girl. You were abducted as a sexual slave during a conflict and now have a young child. You sell fruit on a street stall and earn about $5 a day.

Play the role of ... a newborn baby. Your parents are trying to give you the best future they can. You have food and shelter and people who care about you.

## Who is vulnerable?

▨ Participants line up with their backs against a wall. They think about the role they are playing and take one step forward if their character can agree with the following statements.

1  You have the power to influence people in your community
2  You eat at least two meals a day
3  You receive (received) a primary school education
4  You receive (received) a secondary education
5  You receive gifts or new clothes on religious or national holidays
6  You are listened to in extended family meetings
7  You can afford to meet your basic needs
8  You can afford to see a doctor and buy medicine when you are sick
9  You can afford a place to live
10  You can talk to community leaders about issues affecting your community
11  You drink clean water every day
12  You can read and write
13  You have people who care about you and protect you
14  You are never short of food
15  You are not afraid that people might hurt you
16  You don't have to do dangerous things in order to survive

▨ Now participants should be spread out across the room. Ask each participant to introduce their role, and then ask the others if that person is in the right place for their role, in relation to others (e.g. should the police officer be further forward than the orphaned girl?). **The repositioning and discussion is a very important part of the exercise and enough time must be given for this.**

▨ Ask the participants who have not moved far from the wall what makes them vulnerable. Some of the factors that should be raised are:
• discrimination because of gender, ethnicity, and disability
• the amount of small arms in circulation
• displacement
• lack of law and order
• unsafe or no water sources
• events such as camp closure, demobilisation, ex-combatants returning to their homes
• poverty, and what people must do to earn a living (e.g. need to work in fields or travel to market)

▨ You can ask the participants to adapt their roles to draw out different elements of the discrimination (e.g. would the police officer/president/NGO worker be in a different place if they were a woman?)

▨ You can also ask participants what they would do to decrease the vulnerability of the most vulnerable.

# If you want to make peace, you don't talk to your friends. You talk to your enemies.

Mother Teresa

# Module **2**

## Planning a programme

This module focuses on gathering, analysing, and managing information about protection.

Session 1 introduces participants to a basic protection analysis model using the concepts of threat, protection actors, and vulnerability covered in Module 1.

Session 2 uses scenarios to help participants examine some of the practical and ethical issues involved in gathering reliable information about protection.

Session 3 looks at how to analyse and mitigate risk to beneficiaries, staff, and programmes. The risk analysis is applied to information gathering but is relevant to protection mainstreaming and programming.

# Session 1: Protection analysis

Participants carry out a protection analysis of threats, either based on real-life scenarios in the countries where they are working, or where this is not possible, on pre-set scenarios.

This protection analysis framework is a simplified version of the kind of questionnaire that NGOs would use to gather information from communities and other sources.

This is a key exercise in the training. It applies the concepts from Module 1, begins the Module 2 work on principles for gathering and managing information for protection, and prepares participants to think about responses in Modules 3 and 4.

## Trainer's notes

To deliver this session you will need:

1   to draw the protection analysis template (p. 70) onto flip-chart paper – one for each group

## 50 mins (if using pre-set scenarios)
## 60 mins (if using real-life scenarios)

# Session plan

In this session participants will do a simple protection assessment analysing specific threats.

Explain the protection analysis and run through an example in plenary. Choose the threats the groups will be working on – either a pre-written or a real-life scenario (see below). Split the participants into groups of 4–6 people depending on the number of people and time you have. Give each group a copy of the protection analysis template and get each group to focus on one threat. Each group will complete the protection analysis grid (30 minutes) followed by feedback in plenary (5–10 minutes per group if there are four groups).

There are different ways to run this session depending on the group and the context – either using pre-written scenarios, or asking participants to draw from real-life situations, or a combination of case studies and real-life scenarios.

## Using pre-written scenarios

You can use pre-written scenarios in contexts where it would be too sensitive to use real-life scenarios or control the discussion about them. You could also use these in trainings where participants may not have much common field experience to draw on.

You can also use these with groups who might need to go through some examples before they can come up with suggestions of real-life scenarios.

## Using real-life scenarios

Take 20 minutes at the start of the session to identify and prioritise the threats the groups will work on.

Referring back to the threat categories of violence, coercion, and deprivation from Module 1, ask the participants to identify the main threats to civilians in a conflict or crisis situation. Where possible and appropriate, encourage the group to use real examples from situations they have noticed in the places they are working in or have worked. Examples should not include people's names, but the more specific they can be about a scenario the better the analysis will be.

Write the threats on a flip chart and help the group to prioritise four threats that they will work on, through discussion and/or voting based on the criteria of:
- The impact of the threat
- Gaps in action by other protection actors

Explain that prioritising four threats does not mean that we do not care about the other threats, but that we will prioritise our limited resources by dealing with the threats that have the greatest impact first, taking into account how our work fits with that being done by other actors.

It is important to emphasise that in real life, the analysis of threats would be a participatory exercise done with communities at risk. However, this exercise can be a useful way of doing a preliminary protection analysis with field staff before they go to the field to do an analysis with communities.

Threats identified in this session also form the basis of the objective-setting exercise in Module 4 (p. 133).

## Protection analysis template

| **Threat:** | |
| --- | --- |
| Who is vulnerable to this threat? | Who is responsible for protection and what are they doing? |
| What are the impacts of the threat? | What is the community doing to protect itself? |
| Who is the perpetrator? | What are other NGOs and UN agencies doing? What are the gaps? |
| What might a humanitarian organisation do to fill these gaps? Give at least one example for two different types of activity: advocacy/capacity-building/presence/assistance/voice/information. | |

# Mini case studies

Use a different scenario for each group, writing it out at the top of the flip chart following the template on p. 70.

- While talking to community members during an assessment, you hear that the town chief is charging them a 'tax' in order to access a nearby aid-agency clinic.

- Last night the army came into an IDP camp in which you are working and told everyone that they had four days to return home, after which all supplies and services would be cut off. This is contrary to government policy that says that return must be voluntary.

- Government policy states that all IDPs should receive basic support for six months after their displacement. In the area where you work they only receive one month's food supplies. As a result many are returning home, despite the ongoing conflict in their villages.

- You are working in a rural area with very fertile land, but most of the fields are abandoned. People tell you that they are afraid that if they leave their village to cultivate their fields they will be beaten, abducted, or even killed by the rebels.

- In response to a series of 'insurgent' bomb attacks against civilians, the authorities have imposed travel restrictions that are preventing people from going to work, separating families, and even stopping people from accessing urgent medical care.

- Doctors in a refugee camp close to the border tell you that they are treating many survivors of rape and sexual violence. Most have been attacked while fleeing their homes, but some of the attacks have been in the area surrounding the camp and even inside the camp itself.

- You work in a country where a large number of foreign peacekeepers have been deployed with a mandate to protect civilians. People are telling you that after the peacekeepers search their villages for weapons they have been attacked by rebels. None of the peacekeepers speak the local language.

- Doctors in a clinic in a displacement camp have been told by the local authorities that they cannot treat survivors of rape unless the attack has been reported to the police and a crime report certificate issued.

- A minority ethnic group has been allowed to return to their land after being displaced by conflict, but must pay a 'protection tax' to the warlord to cover the costs of additional security.

# Session 2: Gathering and managing information

Before humanitarian organisations can carry out analysis and response, they need reliable information about the real situation for people facing protection threats. To make the right choices about your work, such as what assistance activities to do or which advocacy strategy to use, you need to keep collecting information, including as part of regular monitoring. However, this information can be hard to gather and can be highly sensitive. This session builds on the protection analysis to identify some of the methods of gathering information about protection and some of the issues that may arise.

This session is made up of two role plays followed by discussion in plenary. The role plays are designed as an optional extra activity to introduce the ideas to groups with limited experience of them, and as an energising activity. For groups where the role plays are not appropriate or necessary, you can use the case-study scenarios to stimulate discussion.

## Trainer's notes

To deliver this session you will need to prepare:

1 copies of the two role-play briefings

2 flip charts and pens

3 copies of Handout 4: Gathering and managing information for protection (p. 77)

## Resource materials for trainers

▨ Handout 4: Gathering and managing information for protection

▨ *Data Collection in Humanitarian Response: A guide for incorporating protection.* Interaction Protection Working Group.

▨ P. Nichols (1991) *Social Survey Methods: A Field Guide for Development Workers*, Development Guidelines no.6, Oxford: Oxfam GB.

▨ R. Scheyvens and D. Storey (2003) *Development Fieldwork: A Practical Guide*, London: Sage.

▨ Shelley Arnstein's ladder of participation can be found at: www.rkpartnership.co.uk/keyconcepts.php

⧖ 90 mins

## Session plan

## Option 1

Tell the participants that in this session we are going to look at some of the key issues involved in gathering information about protection, through two role plays.

The rules for the role play are:
- Everyone must play a part in the role play
- The role play must not be longer than five minutes

Split the participants into two groups, ensuring a gender balance, and give each group a few copies of the role-play briefing (p. 74–5) for their group to read and share in the group. They have ten minutes to prepare the role play.

Ask the groups to perform their role plays, introducing each one before they start. Ask the other participants to think about the key issues in each role play for the discussion afterwards.

After the role plays ask the participants in plenary to comment on what happened in the role plays and link this to their own experience. Use questions to start discussion if necessary, such as 'why do you think the NGO workers didn't gather the information they needed?' and 'how could they have done things differently?'.

## Option 2

If the role plays are not appropriate or necessary for the group, you can divide the participants into groups of four or five and give them ten minutes each to read a scenario and answer the question (p. 75). Each group then has five minutes to read out their scenario and present their responses for discussion in plenary.

## Both options

Pin up three flip charts in different parts of the room, headed: METHODS, DOs, and DON'Ts.

Divide the group into groups of three. Give them ten minutes to use their experience and the issues just raised to agree and note down:
- three METHODS of gathering information about protection
- two things you should DO when you are gathering information about protection
- two things that you should NOT DO when you are gathering information about protection

Give each person a marker pen and ask the groups to split up to send one person to each of the flip charts to write up their answers. They have a maximum of five minutes to do this and go back to their seats. If you have post-it notes you can distribute these instead of marker pens and ask the groups to stick them on the relevant flip chart.

Discuss the three flip charts in plenary and then give a copy of Handout 4 to each of the participants.

## Role play briefing 1

- Read through this sheet
- Decide who will play which role (men can play women and women can play men)

**The rules are:**

Everyone must play a part in the role play, even if only in the background. The role play must not be longer than five minutes.

**Roles:**

**1 or 2 NGO workers**       **1 or 2 police officers**       **Women's group leader**

**Town chief**                    **Women**                         **Town chief's wife**

**The NGO workers**

The NGO workers have heard that women are having problems fetching water and want to find out more about what is happening. They talk to the town chief, police officer/s, and a women's group leader and ask what is happening.

**The town chief**

The town chief knows that some women have been attacked by men in the nearby rubber plantation, but doesn't want to tell the NGO workers about them. If the men find out he has complained about them, they might hurt him or his family. He thinks the best answer is to get the NGO to build a new well for the village so the women don't have to walk past the rubber plantation any more.

**The police officer**

The police officer doesn't know anything about attacks on women. There have been no reports of attacks. This man–woman business is always going on, it is not a police matter. Village women always have lots of boyfriends. The only problem in this village is that he doesn't get paid enough money – not like these fancy NGO workers.

**Women's group leader and women**

When the NGO workers come to interview you, the town chief sends his wife to the meeting to be sure that you don't talk about the rubber plantation. You tell the NGO workers that the only problem is that you don't have a well in your village.

**The town chief's wife**

After the meeting you wait until the NGO workers have gone. You tell the women's group it was good that they didn't mention the men at the rubber plantation. It would only cause problems. You ask one of the women to go and fetch water for your family because it is too dangerous for the town chief's wife to go to the well.

# Role play briefing 2

- Read through this sheet
- Decide who will play which role (men can play women and women can play men)

**The rules are:**

Everyone must play a part in the role play, even if only in the background. The role play must not be longer than five minutes.

**Roles:**

**NGO workers** (one male and one female)

**Women**

**Husband of one woman**

**The NGO workers**

The NGO workers want to start a project to improve women's safety when they are working in the fields. The fields are near an area where there are a lot of ex-combatants, violence, and insecurity. Both NGO workers decide to go to a women's group meeting to talk to the women about what the NGO can do to help them. At the end of this meeting you ask the women if you can come to their next meeting to talk some more.

**Women**

You are having a meeting when two NGO workers come and ask to talk to you about what is happening in your village. You are meeting to organise how to repair the irrigation channel. It is a very important discussion and you must decide tonight. You want the NGO workers to go quickly so you can continue your meeting, but you don't want to be rude to them because they are important people, and they might bring some help to your village.

You ask the NGO workers if they can help repair your irrigation channel. Your village has lots of other needs too: food and clothing, a new school, a clinic, seeds and tools to work the land. Can they help you?

**The husband**

You come home to find that your wife is not at home. You go to the women's meeting and find that your wife is sitting there chatting next to a man from the city. You are very angry. She should be at home looking after you, or working in the fields, not looking for a new boyfriend. You drag your wife out of the meeting and tell her you will beat her for talking to other men.

## Scenarios

- You have been asked to recruit a team of people to carry out a needs assessment with children who were forced to join rebel armies, but have now been demobilised and are going through a rehabilitation process. Many are unaccompanied and have been rejected by their families. What do you need to consider in recruiting, training, and managing your team?

- You go to a village to visit a partner organisation in an area where a number of international NGOs and peacekeepers are based. When visiting a project people tell you about 'foreigners' abusing and exploiting children and young women. What would you do with this information?

- There have been rumours of violence against refugees who have returned to their own country. Your organisation also works in that country and you know that the rumours are incorrect. You want to find out what the refugee community thinks about returning home. How would you do this and how can you also give correct information to the refugees about the situation at home?

- You want to gather information for a humanitarian advocacy campaign about the continual abuses against a minority ethnic group and about how people in power are not protecting them. How would you ensure that the information you gather – including details of abuses and victims – is kept confidential, and that you minimise the risks to people who have given you information?

- You have heard that there are many cases of forced labour in a town where you are working. When you go to visit the local police chief to find out more, he becomes very defensive and tells you that there are no cases of forced labour there. How do you handle this situation and how do you check whether what he has told you is correct?

- Two male members of your humanitarian programme team are visiting a community. You only have one female team member and she is deployed to another programme site 200km away. Another NGO has told you that they have informally heard that there are high levels of sexual and gender-based violence in that community but that they have no evidence. What should you do and who should you talk to?

- You are holding a focus group in a community which is hosting a large number of IDPs. During the focus group everyone agrees that there are no problems between the IDPs and the host community. However, you notice that only one IDP representative has attended the focus group and that he is not speaking much. What should you do during and after the focus group?

# Gathering and managing information for protection

In order to understand and respond to protection threats, we need reliable information about the real situation for people facing them. However, this information can be hard to gather and can be highly sensitive. We need to be sure that we gather information that is useful and that we can trust, and that we are not putting people at further risk when we are gathering or using it.

## WHY do you need the information?

- Be clear about what information you need for your purpose (e.g. assessment, research for advocacy) and that it does not already exist in your organisation or another organisation.

- Make sure that the outcomes you will get from the information are proportionate to the time you spend finding it out.

- Be aware of the information gaps among the humanitarian sector where you are working, and where you can share information with others to mobilise action.

## WHEN should you gather information?

- You will need to gather information in different ways throughout your programme:
  - an initial formal assessment to understand the situation, to do a protection analysis, and to help you plan your approach
  - periodic monitoring visits to see how the situation has changed and how your programme is affecting it and affected by it
  - constant informal observation and informal conversations with communities whenever staff are visiting them or working with them
  - at the evaluation stage of your programme.

- Plan the time and timing before you start any process of formal information gathering as these can take longer than you think. Make sure you have allowed enough time and that your research does not become too burdensome for the community. Be sure to fit it around their schedules and requirements (e.g. market days, farming times).

## WHO do you need to get the information from?

- You will need to speak to a number of different sources to cross-check information to make sure it is more reliable:

  - Identify who can give you the information you need: the government, media, civil society, religious leaders, armed actors, other NGOs, and communities at risk.
  - Decide at the beginning how you will work with the authorities responsible for protection – they will be very important providers of information, but may also feel threatened by your questions.
  - Develop contacts with a range of people and understand what their perspective is and whether they are being influenced or forced by others to say or not say certain things.

☞ | handout 4 |

■ You will need to make sure that the information you gather represents the experience of the people most vulnerable to the threats, not just the most powerful or vocal people in a community:
- Getting information from the most vulnerable people can be difficult. You may have to use structures already in place to reach the most vulnerable people.
- Do a power analysis to help you understand who are the most vulnerable and excluded and who represents their views.
- Be careful about legitimising the views of people who are not good representatives of other people.

## HOW do you gather information (methods)?

■ Decide on one or more ways of collecting information according to what information you need: talking with key informants, site visits, observation, semi-structured interviews, focus groups. Humanitarian organisations have many different tools and resources to guide you in carrying these out.

■ Information gathering (such as interviewing or surveys) means you can also give information about your organisation, the situation, etc.

■ Gather both qualitative and quantitative data. Often it is hard to get precise quantitative data on protection issues. Don't push your informants too hard for this – qualitative and anecdotal evidence can be just as valid, and useful to understand threats and trends.

■ Use secondary sources such as reports by other organisations where they are available and reliable, and work with existing co-ordination mechanisms to share information.

■ You and your colleagues may see or be told about protection threats when you are working in communities and are not conducting a formal information-gathering process. Make sure you have a clear system in place for getting this information listened to and handled properly.

## HOW do you gather information (ethics)?

■ The basic principles are Do No Harm, Confidentiality, Respect, Non-Discrimination.

■ Do a risk assessment before you start. Assess whether speaking to certain people in a certain way about certain issues could increase the threats to the community or to your programme, and how you can avoid this (see Handout 5 – Managing risk).

■ Get informed consent: people must know and understand why you are collecting information and what you will do with it. They must not feel forced to give information and must be able to change their minds about talking to you.

■ Make sure that team members, including interpreters, get the training they need and that they are managed and monitored.

■ When gathering sensitive information (for example on sexual and gender-based violence) make sure that women interview women and men interview men, including interpreters.

■ Keep information anonymous and confidential and be sure that it is stored and shared in a way that prevents it getting into the wrong hands.

## WHAT information should you gather?

▨ Use the protection analysis framework to guide the information you gather about threats, vulnerable groups, perpetrators, protection actors, impacts, and coping mechanisms.

▨ Identify and monitor patterns of violation and abuse and find out when rights have been violated.

▨ Remember there is a distinction between humanitarian organisations and human-rights monitoring organisations. Humanitarian organisations will not usually be seeking to gather individual case studies and testimonies and should avoid doing this if they are not in a position to respond to them or handle the information securely.

▨ Break down (disaggregate) data by gender and other factors such as age or livelihood to identify different vulnerabilities and where, when, and to whom threats are targeted.

▨ Sexual violence is always under-reported and there is always a lack of hard evidence, so anecdotal evidence or observation of behaviour and avoidance tactics can be useful to help you understand its prevalence and trends.

# Session 3: Risk analysis

This session looks at how to assess risk for protection activities, including information gathering. It is equally relevant to modules 3 (mainstreaming) and 4 (protection programming) and the trainer can choose to include it either of these modules. Managing risk will be referred to throughout the workshop and trainers should keep referring back to this session.

## Trainer's notes

To deliver this session you will need:

1  copies of handout 5: Managing risk (p. 83)

2  copies of the Template for feedback (p. 82) for two groups

3  copies of the scenario updates (p. 81) if you choose to deliver them to the groups as 'memos' or 'emails'.

⧖ 60 mins

# Session plan

Explain to groups that it is crucial to assess the risk of your organisation's activities at all stages of your work – from information gathering to response. In contexts with protection needs, there are likely to be issues associated with your work which could put beneficiaries, staff, and humanitarian programming at risk. Organisations need to be able to assess these risks from the outset to make an informed and objective judgement on how and whether to proceed. They also need to be flexible and keep assessing risk to take account of volatile situations. Use Handout 5 to explain the different kinds of risk.

Divide the participants into two groups. Give each group the feedback template (p. 82) and the scenario below. Explain to the groups that they are under pressure in an emergency situation and have ten minutes only to fill in the 'benefits' box and the top line of the last three columns. After ten minutes, interrupt the groups and give them scenario update 1. You can do this by sending an 'envoy' to deliver a 'memo' or 'email' to the group, or by phoning the mobile phone of a member of the group to deliver the message. They have ten minutes to fill in the second line of the last three columns.

Repeat this after ten minutes to deliver the message about scenario update 2. They have ten minutes to fill in the third line of the last three columns. The groups then have ten minutes each to feed back in plenary.

## Scenario

A protection assessment in ten communities where your organisation is delivering a public-health programme. You want to do one focus group in each community and invite women, men, and representatives of the IDP population and minority ethnic groups.

## Scenario update 1:

Update for group 1:
After you have done the first focus group, an IDP tells one of your staff members that they are afraid to come to the meeting because this would anger the host community.

Update for group 2:
After you have done the first focus group a local women's organisation tells you that there was an upsurge in domestic violence after women participated in your focus group.

## Scenario update 2:

Update for group 1:
After you have done the fifth focus group, you receive a phone call from another NGO to tell you that the local authorities have visited their offices and removed all their computers and paper records and appear to be on their way to all the other NGOs.

Update for group 2:
After you have done the fifth focus group, a national staff member who is part of your assessment team who belongs to a certain tribe tells you that they have received death threats from an armed group linked to another tribe.

# Template for feedback

| Proposed activity | Benefits | Risks to beneficiaries, organisation, partners | How might the risks be reduced or managed? | Recommendation |
|---|---|---|---|---|
| A protection assessment in ten communities where your organisation is delivering a public-health programme. You want to do one focus group in each community and invite women, men, and representatives of the IDP population and minority ethnic groups. | | | | |

# Managing risk

Risk can be defined as the relative likelihood of incurring harm – personally or organisationally. It is vital to understand the risk that beneficiaries, staff, and the organisation are exposed to before starting or resuming work.

When analysing risk levels, the evidence base, context, and trends must be clearly articulated. The level of 'real' risk that the organisation and individuals are exposed to must be made clear. We should try to avoid risk where we can, but we must be realistic that nearly all humanitarian environments have a high level of risk – this must be communicated effectively and explicitly and understood by all levels of the organisation. Managers can then make an informed decision and take responsibility for it.

## Risk to staff

Undertaking protection activities in insecure environments could place staff members at risk. This could include being 'accidentally' caught in generalised hostilities, or being deliberately targeted by perpetrators of protection abuses who feel threatened by the staff members' interventions. National and international staff members may face different kinds of risks – international staff may be targeted for perceived wealth or by those opposed to foreign intervention; national staff may be targeted because they are personally known to perpetrators or perceived to be involved in the local conflict.

## Risk to the organisation

Being seen to engage with protection issues could place the organisation (and other humanitarian organisations) at risk. When looking at risk, it is also important to look at the impact that the risks would have on the organisation, and the ability to continue programming. It is not useful to have a list of risks but no understanding of their greater impacts. All known or predicted risks should have narrative explaining the associated level of risk and the impact on the organisation, locally, nationally, and internationally. For example in some countries, doing public advocacy on protection issues could result in the organisation having its operations closed down or being expelled from the country.

## Risk to beneficiaries

In places where protection is a concern, beneficiaries will already be exposed to risks of violence, coercion, and deliberate deprivation. Humanitarian organisations must make sure that their activities do not increase or add to these risks. For example we should make sure that community members who give us information about protection do not face retaliation from perpetrators, or that our activities do not 'displace' the risk onto another part of the community. This should be part of a basic 'safe programming' approach.

## handout 5

### Perceived risk

The level of risk in many instances increases with the acuteness of the needs, and with increased insecurity. This can lead to an assumed acceptance of increased risk without a proper analysis. This is why it is essential to have an ongoing risk analysis to track and measure exposure to risk and how this changes over time.

### Inherent risk

There will always be an inherent risk in working in environments of conflict or in areas where humanitarian organisations are seen as targets. Again, this needs to be clearly acknowledged and where necessary programming needs to be adapted (e.g. by working through local partners to reduce visibility and gain access).

### Acceptable risk

It is impossible to define a global risk threshold where humanitarian operations or protection activities should not go ahead. Every organisation, every programme, and every context is different and decisions have to made on a case by case basis. The final decision rests with senior managers who will have a framework for making this decision. In general, the level of acceptable risk would increase with the direct impact that the organisation was able to have on saving lives and protecting civilians. For instance, running a therapeutic feeding centre in a war zone may well be worth the risk, while small-scale non-lifesaving infrastructure repairs would not.

The matrix should be updated on a regular basis to monitor risk and will require the involvement of mangers who can make informed judgements about activities to be carried out. It may be necessary to determine triggers for stopping activities if risks are realised.

☞ handout 5

## Quick & simple risk assessment framework for protection

| Proposed activity | Anticipated benefits | Possible risks to humanitarian organisation, staff, or beneficiaries | How might the risks be reduced or managed? | Recommendation |
|---|---|---|---|---|
| Protection assessments in ten communities where your organisation is delivering a public-health programme. | Communities will be able to explain to your organisation the patterns of threats facing them, their vulnerabilities, coping mechanisms, and the solutions they wish to see. Your organisation can use this information to improve programming and plan appropriate responses. | Marginalised groups invited to the focus groups will be targeted and 'silenced' by the dominant group. Sensitive information may be 'leaked' to authorities, causing reprisals for the community. National staff may be targeted as spies or informers by rebel groups. | Running separate focus groups or individual interviews for marginalised groups in a private location in the context of e.g. public-health training. Making sure that no names are recorded on the assessment forms and these are passed directly to the programme manager. Clear messaging at all times on your humanitarian mission. | National public-health promotion staff to do a local power analysis and plan focus groups in line with timetable for public-health workshops. Managers to ensure that all staff collecting information are clear on guidelines for information management and system for reporting. All staff to have clear messages on humanitarian principles. Staff who may be particularly targeted to be replaced on assessment team. |
| Give women and girls in IDP camps information about medical, psychological, and legal referral systems for reporting incidents of sexual violence and accessing services. | Women and girls can make informed decisions about how and when to access services available to them and how information about incidents they report will be managed. They can make informed decisions about risk to them and access services necessary for their physical and psychological well-being. | Women face risks when reporting, as information may not be kept confidential and their communities may further victimise them. They may experience reprisals by state officials for reporting incidents. Your organisation is seen to be collecting evidence. In a culture of impunity your organisation cannot be seen to encourage reporting of incidents to authorities that will not take action or may 'punish' those reporting. | Give impartial information so that women and girls can weigh benefits and risk and make informed decisions. Give information through other activities such as public-health committees, women's groups, and FES programme to reduce exposure. Encourage referral systems to provide feedback to communities on how information is kept and used. | Pilot training project in one camp and monitor for three months. NGO should ensure that community mobilisers understand the difference between providing information and encouraging reporting. Develop set 'key messages' for community mobilisers to deliver. Involve community mobilisers in planning and monitoring. |

# For to be free is not merely to cast off one's chains, but to live in a way that respects and enhances the freedom of others.

Nelson Mandela

# Module 3

## Mainstreaming protection

This module looks at how to mainstream protection within humanitarian programmes.
The first session introduces the programme management cycle and the steps that need to be taken at each stage to mainstream protection.

The second session looks at three basic modes of action that can be taken to respond to protection issues that humanitarian staff become aware of during their programming. These are:

- Co-ordinating with or referring to other protection actors

- Local-level advocacy for protection

- Adapting the programme.

In the final session, participants apply this approach to specific scenarios, thinking quickly to decide what they would do in the situation.

# Session 1: Mainstreaming protection

In this module we will look at the basic things that we need to do at each stage of the programme management cycle to mainstream protection into any humanitarian programme.

## Trainer's notes

To deliver this session you will need:

**1** to copy the Programme management cycle grid (p. 90) onto four flip charts

**2** copies of Handout 6: Mainstreaming protection (p. 91)

## Resource materials for trainers

▦ Handout 6: Mainstreaming protection

This session looks specifically at protection using a project cycle approach. It does not cover project cycle management (PCM) in detail. More resources on PCM can be found at:

▦ European Commission PCM Training Handbook
http://europa.int/com/europaid/evaluation/methods/PCM_Train_Handbook_EN-March2002.pdf

▦ The European Commission Directorate General for Humanitarian Aid (ECHO) provides a simplified version: http://europa.eu.int/comm/echo/pdf_files/partnership/pcm_echo_en.pdf

▦ The Department for International Development (DFID) produces a very useful manual of tools: www.dfid.gov.uk/pubs/files/toolsfordevelopment.pdf

Protection mainstreaming needs to be linked with mainstreaming approaches to gender, HIV, and AIDS. Guidelines on these can be found in:

▦ V.M. Walden, M. O'Reilly, and M. Yetter (2007) *Humanitarian Programmes and HIV and AIDS*, Oxford: Oxfam GB, http://publications.oxfam.org.uk/oxfam/add_info_036.asp?TAG=&CID=#contents

▦ *Women, Girls, Boys and Men: Different Needs – Equal Opportunities*. IASC Gender Handbook in Humanitarian Action, www.humanitarianinfo.org/iasc/gender

⧗ 45–60 mins

# Session plan

Tell the participants that mainstreaming protection means taking the safety of civilians into consideration throughout our humanitarian programming no matter what kind of programme it is – public health, livelihoods, shelter, etc. Explain that mainstreaming protection is different from protection programming:

- **Mainstreaming** means applying an approach into the core activities of programmes. It means an organisational commitment to including minimum standards on protection in humanitarian programming. Protection can be mainstreamed into a humanitarian public-health programme by considering civilian safety when we are designing, implementing, and evaluating the public-health activities. The overall objective of the programme remains a public-health objective – but the way we work to achieve this objective means that we take protection considerations into account. Mainstreaming protection in a programme can be a platform for identifying and responding to the need for protection programming.

- **Protection programming** means doing projects which have improving the safety of civilians as their overall objective. We will look at this in more detail in Module 4.

The project cycle management approach helps us to understand the main stages of a programme. Draw the basic diagram of the PCM (from Handout 6 on p. 91) on a flip chart and briefly explain it to the participants.

Divide the group into four, to look at each of the stages: assessment, objective setting, implementation and monitoring, evaluating and learning.

Tell the groups that they are planning a water and sanitation programme for vulnerable households including IDPs and host communities in a region where there is ongoing civil conflict. For their stage in the programme cycle they need to fill in the grid.

They have 20 minutes to write up their answers and five minutes each to present.

Give out Handout 6 and use the key points to sum up the discussion.

## Programme management cycle grid

| Programme cycle stage: | | |
|---|---|---|
| What do you need to think about or do to mainstream protection at this stage? | Who needs to do it? And what knowledge or resources will they need? | What are the constraints/risks? And how will you overcome them? |
| 1 | | |
| 2 | | |
| 3 | | |

# Mainstreaming protection

Mainstreaming protection is different from protection programming:

- **Mainstreaming** means applying an approach into the core activities of programmes. Protection can be mainstreamed into a humanitarian public-health programme by considering civilian safety when we are designing, implementing, and evaluating the public-health activities. The overall objective of the programme remains a public-health objective – i.e. fewer people fall ill or die from cholera – but the way we work to achieve this objective means that we take protection considerations into account. The key things we need to do to mainstream protection at each stage of the programme cycle are summarised in the diagram below.

- **Protection programming** means doing activities which have improving the safety of civilians as their overall objective.

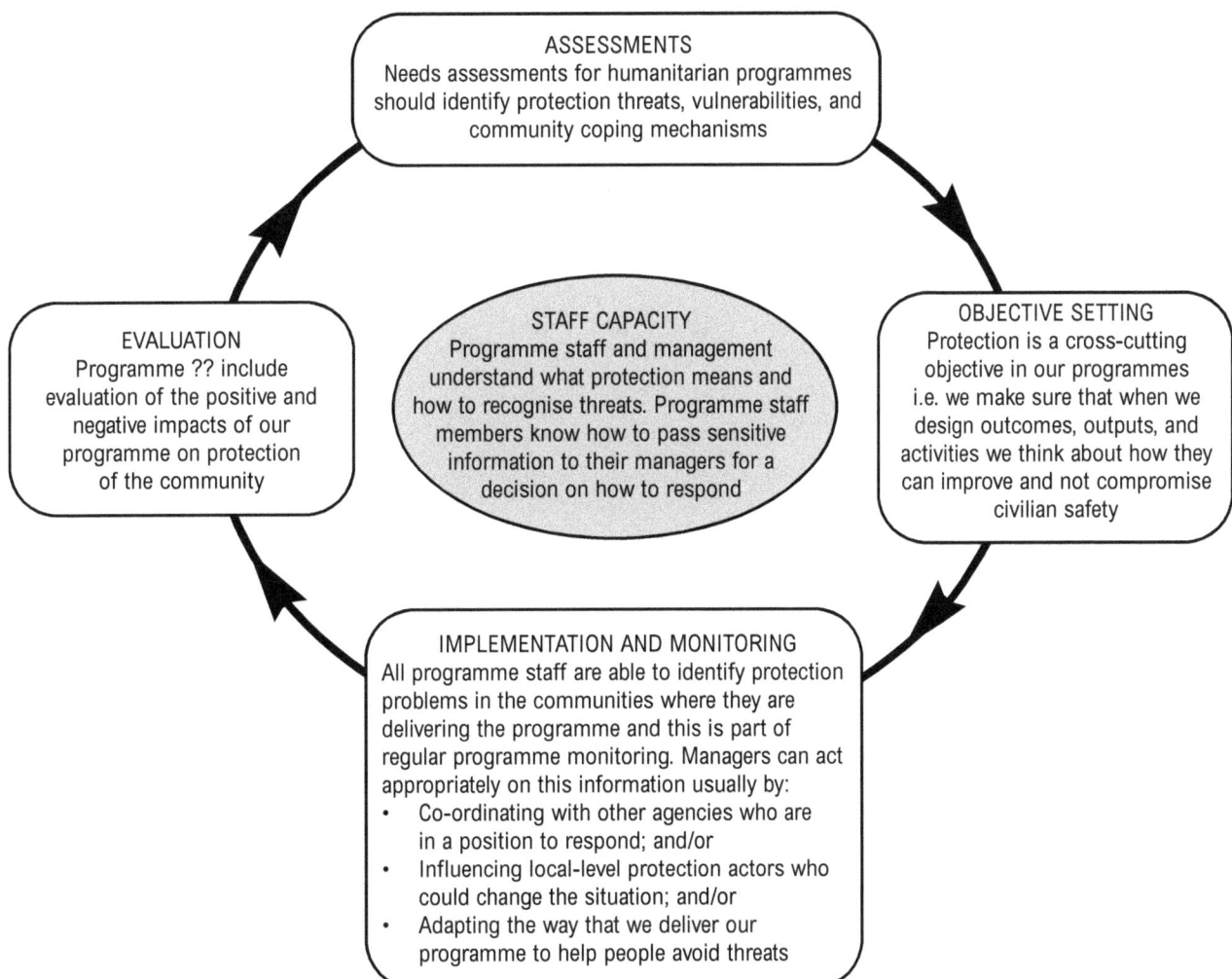

**ASSESSMENTS**
Needs assessments for humanitarian programmes should identify protection threats, vulnerabilities, and community coping mechanisms

**STAFF CAPACITY**
Programme staff and management understand what protection means and how to recognise threats. Programme staff members know how to pass sensitive information to their managers for a decision on how to respond

**OBJECTIVE SETTING**
Protection is a cross-cutting objective in our programmes i.e. we make sure that when we design outcomes, outputs, and activities we think about how they can improve and not compromise civilian safety

**EVALUATION**
Programme ?? include evaluation of the positive and negative impacts of our programme on protection of the community

**IMPLEMENTATION AND MONITORING**
All programme staff are able to identify protection problems in the communities where they are delivering the programme and this is part of regular programme monitoring. Managers can act appropriately on this information usually by:
- Co-ordinating with other agencies who are in a position to respond; and/or
- Influencing local-level protection actors who could change the situation; and/or
- Adapting the way that we deliver our programme to help people avoid threats

☞ **handout 6**

## Further resources on project cycle management

This session looks specifically at protection using a project cycle approach. It does not cover project cycle management in detail. More resources on PCM can be found here:

- European Commission PCM Training Handbook, http://europa.int/com/europaid/evaluation/methods/PCM_Train_Handbook_EN-March2002.pdf

- The European Commission Directorate General for Humanitarian Aid (ECHO) provides a simplified version: http://europa.eu.int/comm/echo/pdf_files/partnership/pcm_echo_en.pdf

- The Department for International Development (DFID) produces a very useful manual of tools: www.dfid.gov.uk/pubs/files/toolsfordevelopment.pdf

Protection mainstreaming needs to be linked with mainstreaming approaches to gender, HIV, and AIDS. Guidelines on these can be found in:

- V.M. Walden, M. O'Reilly, and M. Yetter (2007) *Humanitarian Programmes and HIV and AIDS*, Oxford: Oxfam, http://publications.oxfam.org.uk/oxfam/add_info_036.asp?TAG=&CID=#contents

- *Women, Girls, Boys and Men: Different Needs – Equal Opportunities*. IASC Gender Handbook in Humanitarian Action, www.humanitarianinfo.org/iasc/gender

# Session 2: Options for mainstreaming protection

When humanitarian organisations are implementing and monitoring programmes, they will often identify protection issues that need an immediate response. This session looks at three possible options for response:

- co-ordinating with other agencies

- local-level advocacy

- adapting the programme.

This session is made up of work in three groups followed by discussion in plenary.

## Trainer's notes

To deliver this session you will need:

1 to write the grids (p. 96 and p. 100) onto flip-chart paper for each group

2 to copy the scenarios for each group

3 copies of Handouts 7, 8, and 9 for each participant (pp. 97, 101, 105)

## Resource materials for trainers

- Handout 7: Co-ordination and referral for humanitarian protection

- Handout 8: Adapting the programme to protect

- Handout 9: Local-level advocacy for protection

- On co-ordination: Oxfam International policy compendium note on Humanitarian Coordination, www.oxfam.org.uk/what_we_do/issues/conflict_disasters/downloads/oi_hum_policy_coordination.pdf

- On advocacy: J. Coe and H. Smith (2003) *Action Against Small Arms: A Resource and Training Handbook*, International Alert, Oxfam GB, and Safer World.

## ⧗ 90 mins

## Session plan

Tell participants that in this session they are going to look at what humanitarian agencies can do when they identify a protection issue. Mainstreaming protection often means that the programme does not have a lot of dedicated resources to respond to protection issues – however, staff have a responsibility not to ignore the protection of the people they are working with.

There are many things that humanitarian agencies can do as a minimum response to protection issues when they are mainstreaming protection. These include:

- **Co-ordinating with other agencies:** Working with other agencies so they are aware of the issue and can find a response, e.g. sharing information on protection threats in a particular village in order to identify which organisation can respond.

- **Adapting the programme to keep people safe:** Changing the way we implement our programme to make people less vulnerable, e.g. relocating our latrines so that women do not have to walk through a dangerous place.

- **Local-level advocacy:** Influencing the local authorities or individuals who have the power to reduce the threat, e.g. talking to the local army commander to stop his troops abusing civilians.

Divide the participants into three groups. Each group will look at one of these responses.

Give each group the relevant scenarios and grid on a flip chart. Ideally there should be 6–8 people in each group sitting in a circle or around a table. Give the scenarios to sub-groups of 2–3 people within the group. They have ten minutes to read and think through the scenarios. This will help them think through the information they need to complete the template for feedback. After a few minutes of discussion in pairs they read out the scenario and give their reactions. The rest of the group should suggest other ideas about the scenario.

The participants should then complete the template for feedback on a flip chart and work out how they will present the information back to the larger group. Depending on the level of the group you can choose to give them the handout at this stage to help them to fill in the grid.

Allow a maximum of 15 minutes for each group to give their presentation, and for questions. Presentations should be based on the flip charts they have created, and participants can bring examples from their own experience, act out one of the scenarios, or reflect on their individual learning as they wish. Each group should be encouraged to relate their presentation back to the other parts of the workshop, particularly the modes of action, types of threat, and the roles and responsibilities of NGOs as protection actors. This group then distributes copies of the handout to the other participants.

**Module 3**

# Group 1: Co-ordination with other agencies

## Scenario 1

You are a pubic-health team delivering a cholera response in several villages and IDP camps. When you are talking to one of the communities they tell you about an illegal checkpoint that has been set up by an armed group at a crossroads 10km away, and there are many cases of women being detained and sexually abused. Your organisation is not working in the location where the checkpoint is and is not a specialist in prevention or response to sexual violence. Your agency is a member of the protection cluster which covers the province. What should you do?

## Scenario 2

You are a livelihoods specialist. When you are visiting a community you notice a group of young men and boys carrying heavy building materials. The town chief tells you that they are being made to do forced labour for the army – although this is a major problem, he does not want to complain about it because he is afraid of retaliation. You pass the information to your programme manager who raises it with the UN peacekeepers' human-rights section. They tell her that they cannot do anything about it unless they have precise details of the cases including names and dates. What should you do?

## Scenario 3

Your agency is working to deliver non-food items in a small town. You have seen street children in the town being beaten and abused by the local businessmen and police. Because your agency does not work with children, you passed the information on to an acquaintance who works for a large NGO specialising in child protection. Two months later you see the same thing happening and when you speak to local people they say that nothing has been done. What should you do to follow this up?

# Template for feedback

1. How do you identify which agencies can respond to a particular protection problem?

2. In the place where you are working, which agency could you co-ordinate with to respond to cases of:

   a) Recruitment of child soldiers

   b) Arbitrary arrests and detention

   c) Sexual violence

3. What could you do if there are no agencies who can respond to the problem?

4. How can you make sure that the information that you pass on does not put people at more risk?

5. What follow-up should you do to make sure that action has been taken?

# Co-ordination and referral for humanitarian protection

It is not possible for a single organisation to deal with all the protection needs in a country or often even within a community. By co-ordinating and co-operating, different humanitarian actors can together provide a more comprehensive response, using their different areas of expertise, their different levels and kinds of resources, and their different geographic locations of operation.

If you know exactly which agency has the expertise and resources to respond to the specific protection need you have identified you can directly refer it to them. However, sometimes you may not know which agency is the right one to go to, or the issue may concern several different agencies. In this case, you can use humanitarian co-ordination mechanisms.

## The protection cluster

▨ In many humanitarian settings, the 'cluster system' will be operational. This system was created by the UN in 2005 to meet the need for better co-ordination and to address gaps in humanitarian response around nine key areas of humanitarian activity, including protection. The cluster system was piloted in several countries and should be applied in all major new disasters.

▨ UNHCR is the global lead for the Protection Cluster and chairs the Protection Cluster in countries where there is significant displacement. In countries where there is not much displacement, OHCHR, UNICEF, or UNHCR decide who will lead the Protection Cluster. The lead agency is the agency of 'last resort' – this means that if there is a protection need that is not being addressed, it is ultimately up to the lead agency to make sure the gap is filled.

▨ At a field level, the cluster will meet regularly to share information on protection issues, identify gaps in information and response, and work together to take action to respond to them. The cluster should bring together the UN agencies and humanitarian NGOs active in protection, and it is usually managers or people with specific protection responsibility who attend on behalf of their organisation. Representatives of state structures responsible for protection should also attend. In some locations there will also be representatives of or links to national and local NGOs. Representatives of the cluster will liaise with state authorities and UN military where necessary on protection.

▨ In some countries, the cluster lead may decide to set up sub-clusters – smaller groups which deal with a specific kind of protection need or a specific geographic area, or who are ready to react rapidly to new crises.

☞ **handout 7**

## Other humanitarian co-ordination mechanisms

In most humanitarian settings there are other co-ordination mechanisms where representatives of agencies regularly meet to share information and co-ordinate response. These include security meetings, humanitarian information meetings (often co-ordinated by UN OCHA), and NGO heads of agency meetings.

## Remember:

- Identify who you will co-ordinate with carefully according to who has the practical means or the mandated authority to respond.

- Handle information sensitively. Make sure that the information is passed on in a controlled way that will not put people at further risk.

- Do not collect detailed information about individual cases unless you and your organisation have the skills, experience, and systems to do so safely, confidentially, and sensitively, and are able to take action. Tell a specialist organisation about any information you have so that they can take further action.

- Give communities information about where they can go to access other agencies directly.

- Identify the right person within your organisation and within the organisation you are co-ordinating with.

- Follow up periodically to make sure action has been taken.

- Check with your managers and tell them what you are doing as they may also have links with the same organisation.

- Referral means passing information about protection needs to other agencies who are in a position to take practical action. It does not include passing cases to the police or national or international criminal courts.

## More information:

Oxfam International policy compendium note on humanitarian co-ordination, www.oxfam.org.uk/what_we_do/issues/conflict_disasters/downloads/oi_hum_policy_coordination.pdf

# Group 2: Adapting the programme

## Scenario 1

You are implementing a programme building latrines in a refugee camp. You have built 50 of the 100 latrines and are under a lot of time pressure to finish the programme to meet the donor's deadline. When you are talking to the women in the sanitation committee as part of a monitoring visit, they tell you that they are worried about the location of the latrines as there have been some recent cases of rape there. What should you do?

## Scenario 2

You are in the exit phase of programme to deliver clean water to a village. You are about to remove a water tank from one area of town where an ethnic minority lives, because there are now enough wells to meet everyone's requirements. However, the chief from this area of town asks you not to remove the bladder as they are afraid of abuse from the dominant ethnic group if they have to share the well with them. What should you do?

## Scenario 3

You are delivering non-food items to IDPs who are just about to return home. On a recent distribution the IDPs seemed reluctant to receive the goods. When you ask the IDP leader about this, he tells you that in the time between receiving the goods and travelling home, they are targets for armed robbery. What should you do?

# Template for feedback

What system would you need to have in place to work out if you need to adapt your programme to improve people's safety?

What information do you need to help you decide how to adapt the programme, and where can you find this information?

Who in your organisation should you pass the information to about the need to adapt the programme? Who is in a position to make a decision to adapt the programme?

What are the three main constraints in adapting a programme and how can you deal with these?

1

2

3

# Adapting the programme to protect

Humanitarian situations can be fast-changing and unpredictable, therefore programmes will inevitably need to be adapted during implementation. Not only do external events change and evolve, but we also continually develop our own understanding of the situation and learn while we implement. We need to constantly monitor the external situation and our programme to:

1   check that our overall objectives remain relevant and appropriate

2   adapt our activities as appropriate.

## Monitoring changes in the context

Understanding the protection context must be part of the analysis that informs the design of the programme. However, this analysis needs regular review and updating; a programme designed at one point in time can quickly become irrelevant to real needs as the situation changes. For example, an escalation in fighting or new patterns of displacement will change the protection and assistance needs of a community and mean we need to change the way we deliver our programme.

Senior managers should be responsible for ensuring that the context analysis is updated. This can be done through regular field monitoring reports and specific review meetings or workshops at appropriate intervals. More informally, field staff need to be constantly telling their managers about the changes that they see and hear in communities. Humanitarian organisations should also work together to share information and analysis rather than duplicating each other's efforts.

## Monitoring the programme implementation

Programme monitoring tends to focus on whether the planned activities have been delivered and the results are being achieved. For example, if we planned to build 20 latrines we will monitor whether we built 20, whether the quality is good, people use them, like them, and that they are kept clean. However, a good protection mainstreamed humanitarian programme should not only be monitoring pre-set technical details but should also register people's perceptions of their own safety while using the facilities and services provided. If we are providing clean water, the technical team should monitor water quality and should also make sure the water point does not cause problems between host communities and IDPs and that women are not abused or attacked when collecting water.

Staff and communities need to be aware of the need for analysis and sharing of results. Community monitors may all be aware of women being attacked in the latrines at night, but unless they know that the agency is prepared to do something about it, they may assume that their monitoring duties are limited to checking that latrines are being used and maintained.

In a rapid-onset emergency, a real-time evaluation is an opportunity to review the response and to identify lessons for future learning as well as areas for programme change. This process is led by an outsider to the programme who, through a series of interviews and field visits, enables the staff to step back and critically appraise the programme. The evaluation is usually carried out around six weeks after onset, giving the programme manager ample time to make the necessary changes.

☞ **handout 8**

## Decision-making and adapting programmes

Identifying the need for change is only the first step – actually bringing about that change is often difficult. The change could be quite small, for example adding locks to the design of latrines if women tell us that they would otherwise be scared to used them. However, the change might be quite radical, for example we might have to relocate our programme if new threats are displacing the beneficiaries, or be prepared to add in new activities to reduce people's vulnerability.

Strong and confident leadership is necessary to make changes and adapt. Needing to change is not a sign of bad planning but rather a reflection of the reality of the humanitarian response. Donors are usually open to changes that can be justified by a good analysis. It is important to invest in an ongoing sharing of analysis and good communication with donors so that they understand the rationale for change and can advise you on their internal processes.

Before any changes are made, the risk assessment will have to be updated to ensure that beneficiaries and staff will not be put in further or different danger. This will use the context analysis to ensure that any possible risks of the planned change are carefully considered. And of course, once the change has been decided and implemented, the monitoring will need to continue.

## Resources:

- O. Bakewell (2004) *Sharpening the Development Process: A Practical Guide to Monitoring and Evaluation*, Praxis Guide no. 1, Oxford: INTRAC, www.intrac.org/publications.php?id=31

- *Save the Children Toolkit: A Practical Guide to Assessment, Review and Evaluation*, www.savethechildren.org.uk/en/54_2359.htm

- Emergency Capacity Building Project (2007) *Impact Measurement and Accountability in Emergencies: The Good Enough Guide*, Oxford: Oxfam GB, http://publications.oxfam.org.uk/oxfam/display.asp?K=_2006111410173391

# Group 3: Local-level advocacy

## Scenario 1

You are working in a community when you hear that a new wave of IDPs arriving in a nearby location are being beaten by the local army battalion who are taking orders from a regional commander. There is an international peacekeeping mission in the area but they are not deployed to the area. Who should your advocacy target be and how do you decide?

## Scenario 2

From conversations between communities and your local staff you have identified that the local governor is responsible for blocking aid to a village populated by a minority ethnic group. You know that he is very reasonable in his discussions with NGOs but has control over armed groups who can make things very difficult for your programme and your beneficiaries. How do you decide what influencing tactic to use?

## Scenario 3

You have finally been granted a meeting with the district commissioner. You have been trying to contact him for some time, to persuade him to take action on the illegal checkpoints in the area, where civilians are being threatened. You know of a lot of such cases from your team's observation and from what beneficiaries have told you. How should you prepare for the meeting?

## Template for feedback

| How do you decide on an advocacy target? | What are the advantages and risks of different types of advocacy? |
|---|---|
| | Persuasion |
| | Mobilisation |
| | Denunciation |
| What information do you need before you start your advocacy? | |
| | |

☞ **handout 9**

# Local-level advocacy for protection

## What is local-level advocacy for protection?

Advocacy is influencing those with power to change policies (laws, norms, and official rules etc.) and practices (the way in which these policies are or aren't applied) to make a positive difference in the lives of beneficiaries. Local-level advocacy for protection means influencing those with power in the places where we are programming either to better protect or to stop abusing civilians.

## Who do we influence?

The people we are trying to influence are called 'advocacy targets'. We need a good understanding of the local context in order to identify who to target. Where we do not have this information ourselves, we can work with others to do a power analysis.

There are four types of advocacy targets for local-level advocacy on protection:

1.  Individuals/organisations who have the responsibility to protect civilians
    (e.g. the police, army, local authorities, peacekeeping missions)
2.  Individuals/organisations who can influence those who have responsibility to protect
    (e.g. the mayor, UN civil-military staff)
3.  Individuals/groups who are perpetrating the protection threats (e.g. militia leaders, local authorities)
4.  Those who can influence the perpetrators (military prosecutors, inter-community committees)

The same target may fit into different categories depending on the issue, as in many cases those responsible for protection are also perpetrators of protection threats. Some common advocacy targets for local-level protection advocacy might include:

▨ Traditional authorities (local village chiefs, the inter-ethnic community platforms, etc.)

▨ Local civilian authorities (local administrators, the mayor, the district commissioner, the governor, the provincial assembly, etc.)

▨ Local police and other law enforcement (local police commanders, the provincial police chief, local customs or tax authorities, etc.)

▨ Military authorities (local military commander, military prosecutors, regional military commander, etc.)

▨ Peacekeeping missions' military officials (local military commanders, civil–military relations officials or humanitarian affairs officers who attend NGO co-ordination fora, etc.)

▨ Peacekeeping missions' civilian staff (child-protection section, human-rights section, rule-of-law section, etc.)

If we know our advocacy target personally (or can easily contact them), direct interaction is usually the best way of influencing them.

☞ **handout 9**

If our advocacy target is harder to reach (because they are far away, or very senior) we can try to influence them by contacting others who can transmit our messages to them. These people who are close to our advocacy targets are usually called 'secondary' targets. We can influence high-level targets, for example the regional military commander, by using secondary targets such as a local battalion commander, a peacekeeping mission's commander who works with the army locally, or a peacekeeping mission's civilian staff member who is responsible for protection-related issues.

## What are the different types of advocacy?

There are three types of advocacy:

▪ **Persuasion:** talking privately to convince your advocacy target to change its policy or practice

▪ **Mobilisation:** telling other organisations or countries what is happening so that they can influence the authorities or perpetrators to change their policy or practice

▪ **Denunciation:** making public what is happening in order to put pressure on your advocacy target to change its policy or practice.

## How can we influence protection advocacy targets?

▪ **Meeting and talking:** setting out our concerns and recommendations on the telephone, in face-to-face conversations, or in cluster and other co-ordination meetings. Prepare your arguments and know in advance what you want to achieve. Since this type of influencing is often quite informal, this kind of local-level advocacy does not normally require manager approval or sign off. Don't forget that sometimes it can be better to share sensitive or confidential information in private meetings rather than large public gatherings. If you are unsure about which type of information to share in meetings, consult your manager.

▪ **Writing:** sending an email or a letter, or providing the target with a briefing note (facts, figures, or arguments that back up our demands) or other factual documentation (programme information, public or private research, media reports, etc.) Because other written communications like official letters or new briefing notes tend to be more formal, it's usually a good idea to keep your manager informed of these exchanges (and if necessary, check the content with them in advance).

You could also do advocacy through media, awareness-raising through training, employing publicity and marketing tactics, and community mobilisation. These are not normally things we do as part of local-level advocacy when we are mainstreaming protection as they require significant time and resources.

☞ **handout 9**

# When you are doing local-level advocacy on protection, DO:

- Make sure you have reliable information about the problem

- Prepare your argument well and be clear about what change you are looking for

- Identify your target(s): have a clear understanding of who has the power to resolve the problem or to influence those who can

- Be calm, polite, and diplomatic when approaching your advocacy targets

- Think about the risk of approaching a particular target on a particular issue – carry out a risk assessment

- Co-ordinate and be aware of what other agencies are saying on the same issue

- Plan tenacious follow-up – either be prepared to do this yourself or ask colleagues to help you

- Record what you do and what the follow-up is and share with colleagues and allies.

# When you are doing local-level advocacy on protection, DON'T:

- Act on rumours

- Share confidential or sensitive information or use names of individuals who have given you information

- Be over-confrontational or jump straight to denunciation – this will limit your space to influence

- Do advocacy that would put your programme, beneficiaries, or other agencies at risk

- Let advocacy drop if you don't see a result straight away.

## Resources on advocacy for humanitarian protection:

J. Coe and H. Smith (2003) *Action Against Small Arms: A Resource and Training Handbook*, International Alert, Oxfam GB, and Safer World.

# Session 3: What would you do?

Participants are given scenarios and asked to discuss what they would do in each situation.
This session can also be used as a core exercise at any point in the training. See p. 143, Core exercises.

## Trainer's notes

To deliver this session you will need:

1  the 'what would you do…' cards (provided on pp. 110–14 and available in colour at the back of the book)

## Resource materials for trainers

No answers are provided for the scenarios on the 'what would you do…' cards. The training team should be prepared to help participants think through appropriate actions and to guide them towards the relevant tools, standards, and principles.

⧗ 20–60 mins

# Session plan

Each 'what would you do…' card presents a short scenario that participants may come across in real life. The cards can be used in various ways. Here are some suggestions:

1   Split participants into groups and give each group a card to discuss. Ask for suggested actions using each mode of action (persuasion, mobilisation, denunciation, capacity-building, substitution). Feed back in plenary.

2   Split participants into groups and give each group a card to discuss. Ask for actions that can be taken:

   **a** to respond immediately to prevent a violation or alienate its effects

   **b** to restore a victim's dignity through rehabilitation, compensation, or repair

   **c** to build an environment that addresses the causes of the threat and vulnerability over the longer term.

3   In plenary ask a participant to choose a card and read it out. The whole group discusses the issue.

4   In plenary give cards out to groups of 2–3 participants and ask them to discuss and feed back.

5   Set up circles of chairs facing each other (same number of chairs in inner and outer circles). Each circle should have 4–5 chairs (8–10 participants). If you have more participants, create sets of circles. The participants in the inner circle are advisers. The participants in the outer circle each take a card, and ask the adviser opposite for help in dealing with the problem. After 2–3 minutes the participants in the outer circle move onto the next chair and present the same problem to the next adviser. Afterwards in plenary ask people to discuss the advice they were given or gave, and what they found particularly challenging about the scenarios they discussed.

# 'What would you do?' cards

| | |
|---|---|
| **What would you do**<br><br>**...if a community asks you to visit them more often as it makes them feel safer?** | **What would you do**<br><br>**...if you hear about a woman having sex with up to 20 UN staff a night in order to feed her extended family?** |
| **What would you do**<br><br>**...if you go to monitor water quality at a well and see many young men hanging around, and the villagers appear to be frightened and intimidated by them?** | **What would you do**<br><br>**...if soldiers will only allow you to take medicines through a checkpoint if you leave them enough supplies for their own clinic?** |

**What would you do**

...if you are told that a family in a village has a large store of weapons in their house?

**What would you do**

...if a school which you helped rehabilitate reopens, but children from one ethnic group have been removed from the register?

**What would you do**

...if a town chief refuses to allow you to carry out an assessment in one of his villages even though the villagers have asked for your help?

**What would you do**

...if a community at risk of famine refuses to accept food aid because it is afraid of attacks by rebels who want to steal the food?

**What would you do**

...if you hear from a community that one of their newly appointed police officers committed abuses against civilians in a recent war?

**What would you do**

...if you report attacks on civilians at a protection co-ordination meeting but no action is taken and the situation is getting worse?

**What would you do**

...if women who report sexual assaults to the local police are arrested for adultery?

**What would you do**

...if you have carried out an awareness-raising campaign on human rights with a community, but this has not stopped attacks by rebels?

**What would you do**

**...if the main reason people are hungry is because they cannot travel to market without identity papers, but the government has stopped issuing such papers?**

**What would you do**

**... if you go to buy vegetables in the market and find hardly any traders selling goods even though there was a very good harvest?**

**What would you do**

**...if you suspect that men and young boys are being sexually abused while being illegally 'arrested', but no one will talk about it?**

**What would you do**

**...if you go to monitor a programme at a health centre and find it has been abandoned?**

What would you do

**...if the area governor tells NGOs they are not allowed to employ displaced people and refugees on their staff?**

What would you do

**...if you see a group of frightened looking civilians being held back as you are waved through a checkpoint?**

What would you do

**...if all the women disappear as you drive into a displacement camp?**

What would you do

**...if you help refugees newly arrived from a war zone climb off trucks and buses, but notice that there are no men or boys among them?**

# Module 4

## Programming for protection

This module looks at programming options to address two common types of protection threats:

■ reducing the risk of sexual violence

■ supporting durable solutions to displacement.

Participants then apply these ideas to create their own logical frameworks for other protection programmes designing objectives, outcomes, activities, indicators, and sources of verification.

# Be the change you want to see in the world.

Mahatma Gandhi

# Session 1: Options for protection programming

This session introduces some of the options for protection programming, giving some practical examples on major protection themes, before the groups design their own protection programmes in the next session. Facilitators can choose to omit this session and go directly to Session 2 (p. 133).

## Trainer's notes

To deliver this session you will need:

1  to prepare the feedback grids for each group

2  the programming options cards for each group (provided on p. 124 and p. 132 and available in colour in the back of the book)

3  copies of Handouts 10 and 11 (pp. 122, 129).

## Resource materials for trainers

▨ Handout 10: Sexual violence

▨ Handout 11: Supporting durable solutions to displacement

⏳ **90 mins**

## Session plan

Explain to participants the distinction between protection programming and mainstreaming protection. Protection programming means doing protection activities which have improving the safety of civilians as their main aim. All humanitarian programmes should be mainstreaming protection, but not all will be working in a context where there is a need or a possibility for them to implement protection programmes.

The decision whether to do protection programming should be informed by the protection assessment, which will identify what the problems are, where the gaps are in response, and what solutions communities would like to see.

This module will look at different options for protection programming. Remind participants of the main modes of action for protection that were outlined in Module 1:

- To reduce the threat: advocacy, capacity-building, presence.
- To reduce vulnerability: assistance, voice, information.

This session will apply these to two issues which are common to nearly all humanitarian crises – displacement and sexual violence.

Divide the participants into two groups. Participants should choose whether to be in the sexual violence or the displacement group depending on their own interest, experience and expertise. Ideally participants should be split evenly between the groups and each group should be representative of the participants as a whole.

Explain that each group will feed back in plenary. Feedback will cover the basic issues associated with each topic or area of work, and offer the possibility for further discussion. Every participant will receive a copy of all the handouts at the end of the module.

Follow the instructions for each group.

# Group 1: Sexual violence

This session is made up of small-group work based on a case study followed by a presentation in plenary.

## Trainer's notes

To deliver this session you will need:

1 to copy the case study on p. 123 for each group member

2 to draw the template for feedback on p. 121 on a flip chart

3 to copy Handout 10 for each participant in the workshop

4 to draw the protection analysis template from p. 68 on a flip chart

5 to copy the programming options cards on p. 124 and cut them out

6 to provide flip-chart paper and pens

Ideally there should be 4–8 people in this group sitting around a table or in a circle.

## Resource materials for trainers:

- Handout 10: Sexual violence

- RHRC consortium: 'Gender-based violence tools manual for programme assessment and programme design, monitoring and evaluation in conflict-affected settings', www.rhrc.org/resources/gbv/gbvtools/manualtoc.html

- Inter-agency Standing Committee, 'Guidance on Gender-based Violence Interventions in Humanitarian Settings', www.humanitarianinfo.org/iasc/content/subsidi/tf_gender/gbv.asp (also available in hard copy)

- UNHCR, 'Sexual and Gender-based Violence Against Refugees, Returnees and Internally Displaced Persons: Guidelines for Prevention and Response', www.rhrc.org/resources/gbv/ gl_sgbv03.html (also available in hard copy)

- The World Health Organisation provides online technical guidance on sexual and gender-based violence at: www.who.int/hac/techguidance/pht/SGBV/en/index.html and specifically on ethics of monitoring sexual violence: www.who.int/gender/documents/ethicssafety_web.pdf

## Session plan

## Brief the group:

Sexual violence (including rape, attempted rape, forced prostitution, forced pregnancy or abortion, sexual harassment or abuse, and sexual exploitation) is a feature of all conflicts and is often used as a deliberate tactic in conflict. Given the lack of reporting of such abuse, recent guidelines advise all humanitarian actors to plan on the basis that sexual violence is taking place regardless of whether evidence is available or not.

In this exercise the group members will carry out a protection analysis (see p. 68) of a case study and examine a range of programming options to assess the priority actions that can be taken to reduce the risk of sexual violence. Although the case study in this exercise looks at widespread and systematic sexual violence in conflict, the facilitator should emphasise that sexual and gender-based violence can occur at all levels of society (within families, communities, schools, and workplaces, etc.) and includes sexual exploitation by aid workers.

## The group work:

The facilitator should give out Handout 10 and talk through it with the participants. They will need to understand this information in order to complete the group work. The group members should now complete the template for feedback, which will form part of their feedback presentation.

Give out the case study and allow the group members enough time to read it through. The group members should then carry out a protection analysis using the template you have provided.

Once the protection analysis is complete, tell the participants that they are part of an inter-agency working group leading the humanitarian response to the situation. They have been asked to recommend actions to reduce the risk of sexual violence by both reducing (1) the level of threat and (2) the vulnerability of women and girls to that threat.

Give out the programming options cards. Tell participants that they have enough collective resources to carry out five immediate actions. They should use the protection analysis to help them rank the priority actions, taking into account the impact of the threats and the actions of other humanitarian actors. They can also propose their own programming options in addition to cards they have been given.

In plenary the group should structure their feedback as follows:

1   A brief presentation on sexual violence using the feedback template provided (5 mins)
2   A summary of the situation described in the case study using the threat analysis format (5 mins)
3   An explanation of the five actions chosen and why they have chosen these options (10 mins)

During feedback, group members should be encouraged to be concise and to the point, and to make their feedback interactive and lively.

At the end of the presentation the group members should give Handout 10 to all other participants in the workshop for future reference. Allow time for questions.

## Template for feedback

| What is sexual violence? | What are the causes of sexual violence? |
|---|---|
| Definition:<br><br><br><br><br>Give some examples: | |
| What are the impacts of sexual violence on (1) those who directly experience violence and (2) the wider community? | What are the blocks to reporting sexual violence? |
| | |

☞ **handout 10**

# Sexual violence

Sexual violence is a form of gender-based violence. Gender-based violence (GBV) is a term used to describe a wide range of harmful acts committed against a person on the basis of their gender, including sexual violence, domestic violence, female-genital mutilation, forced early marriage, and widow killings. Sexual violence (including rape, attempted rape, forced prostitution, forced pregnancy or abortion, sexual harassment or abuse, and sexual exploitation) is a feature of all conflicts and is increasingly used as a deliberate tactic in conflict.

Sexual violence is a violation of universal human rights and can be a crime against humanity. It is most often targeted at women and girls, but can also affect men and boys, particularly when they are in detention.

In emergency situations sexual violence can be one of the most immediate and dangerous forms of GBV. Sexual violence is under-reported everywhere and almost impossible to measure accurately in emergency settings. Recent guidelines by the Inter-Agency Standing Committee advise aid agencies to assume that most forms of GBV, and particularly sexual violence, are taking place in any situation regardless of whether evidence is available.

Evidence of sexual violence can be very hard to gather, as people are often unwilling to report attacks for fear of stigmatisation, further victimisation, or reprisals. Under-reporting is common when survivors may be rejected by their family and community, be forced to marry their attacker, be themselves prosecuted, or where people do not trust the authorities to protect their identity, carry out an investigation, and prosecute the perpetrator/s.

Discrimination and power imbalances are the root causes of all GBV. The breakdown of the normal protective structure during conflict and crisis (police, judiciary, family, and community structures) can leave women and girls more vulnerable to attack. Events such as displacement, separation from parents and other family members, detention, reliance on humanitarian aid and so on, place people at risk. The perpetrators of sexual violence are often those in positions of power and authority at all levels: within the family and community including teachers, community leaders, and aid workers; as well as armed actors in conflict who may be committing acts of sexual violence opportunistically or as part of a military strategy. Perpetrators can be directly responsible for the violence or indirectly responsible for assisting it to happen.

The impact of sexual violence can include physical injury, pregnancy, miscarriage, gynaecological and other health complications, or sexually transmitted diseases including HIV, and it can also have a severe psychological impact. Humanitarian actors are recommended to focus both on preventing sexual violence, and on responding to the needs of survivors. The Inter-agency Standing Committee has produced guidelines on responding to sexual and gender-based violence that give detailed guidance and are recommended reading for all participants in this workshop.

## Resources:

- Inter-agency Standing Committee, 'Guidance on Gender-based Violence Interventions in Humanitarian Settings', www.humanitarianinfo.org/iasc/content/subsidi/tf_gender/gbv.asp (also available in hard copy)

- UNHCR, 'Sexual and Gender-based Violence against Refugees, Returnees and Internally Displaced Persons: Guidelines for Prevention and Response', www.rhrc.org/resources/gbv/gl_sgbv03.html (also available in hard copy)

- The World Health Organisation provides online technical guidance on sexual and gender-based violence at: www.who.int/hac/techguidance/pht/SGBV/en/index.html

# Case study: sexual violence in conflict

Villagers from an ethnic minority associated with anti-government rebels have been attacked by militias believed to be supported by the government. In some cases entire villages have been destroyed and villagers have moved into displacement camps around larger towns. All community members are at risk of violent attacks, with women and girls being targeted for rape and other sexual assaults in a systematic manner. They are most at risk when leaving their villages or camps to collect water or firewood. Often these attacks include kidnapping for a number of days, during which women and girls are gang-raped and subjected to other violence (beatings, whipping, etc.). Women and girls are very reluctant to report attacks as they have no faith in the police force or judicial system. The police claim they have insufficient resources to provide security in displacement camps or villages. There is a small international peacekeeping force that would like to provide more security, but is under-resourced.

There are very limited medical services in villages, which are mostly serviced by aid-agency mobile clinics. The regional governor has announced that medical treatment for survivors of sexual violence can only be given once the attack has been reported to the police and a case number allocated, in order to encourage reporting. Medical staff who provide treatment to people who have not reported attacks to the police are subject to police investigation and some have been arrested themselves.

There is little discussion of sexual violence within the community. In the past some rape survivors were forced to marry the rapist to avoid bringing shame on their family. Sexual violence within families and communities was not acknowledged or reported prior to this conflict, and families generally 'settled' rape cases among themselves. Rape survivors are stigmatised and may be subject to further victimisation by their own family and community. However, the sheer scale of sexual violence directed at women and girls has created a greater awareness of the issue, and there are indications that some community leaders are starting to speak out about it.

Some women and girls who have been attacked are now pregnant or have given birth, and many suffer from gynaecological complications and other medical problems due to the brutality of the attacks. There is no data available, but it is assumed that women and girls who have been attacked are at risk of contracting sexually transmitted diseases, including HIV.

Aid agencies are in the process of establishing water supplies within camps, but this is not possible in the villages because they are spread out over a very large area. Women and girls sometimes collect water and fuel in groups to minimise risk, although some feel that this makes them too visible a target and continue to go out alone. Families are increasingly sending older women and younger children in the hope that they are less likely to be targeted. Fuel is available for sale at local markets, but is expensive.

# Programming options cards

| | | | |
|---|---|---|---|
| **Provide water sources in safe areas (camps only)** | **Train humanitarian workers on sexual and gender-based violence** | **Persuade peacekeepers to alter their patrol patterns in camps and villages** | **Lobby for a larger peacekeeping force and a stronger mandate** |
| **Lobby the governor to allow treatment of survivors without a police report** | **Review your organisation's programmes to ensure that they do not put people at further risk** | **Identify perpetrators and inform them that what they are doing is illegal** | **Collect data on sexual violence and establish a data storage system** |
| **Develop a standard reporting format for agencies to provide data on sexual violence** | **Train your own staff and partners on the Red Cross Code of Conduct especially regarding sexual exploitation** | **Persuade the judiciary to punish perpetrators** | **Set up a confidential follow-up system of medical and psychological support for survivors of attacks** |
| **Set up inter-agency focus groups with men and boys to discuss sexual violence and its impact** | **Research and publish a report denouncing the government**<br><br>NB: if you choose this option, you cannot implement any other activities as you will be expelled from the country | **Set up inter-agency focus groups for women and girls to discuss the threats they face and actions they want aid agencies to take** | **Provide fuel to women in camps**<br><br>NB: you can only do this for three months, and you can only choose ONE other activity as this will use up almost all your immediate resources |

# Alternative session

An alternative version of this exercise looking at different aspects of sexual violence can be carried out using the scenarios below. The additional resources listed on Handout 10 also contain exercises and materials focusing on a broader range of sexual and gender-based violence issues.

## The group work:

The facilitator should give out Handout 10 and the participants should complete the template for feedback on sexual violence, which will form part of their feedback presentation.

Give out the scenarios to individuals or pairs to read through. After a few minutes each person or pair reads out their scenario to the group and explains what they would do, or asks the group for advice.

You may find it useful to photocopy the programming options cards for participants.

In plenary the group should structure their feedback as follows:

1 A brief presentation on sexual violence using the template provided (5 mins)
2 Each individual or pair reads out their scenario and explains what they would do in that situation.

Allow time for questions and comments after each scenario, but the facilitator should manage time to ensure that all scenarios are covered in this session.

During feedback, group members should be encouraged to be brief and to make their feedback interactive and lively. At the end of the presentation the group members should give the handout to all other participants in the workshop for future reference. Allow time for questions.

## Scenarios: reducing the risk of sexual and gender-based violence

1 One of your partner organisations has asked for support in raising awareness about sexual exploitation of women and children in a refugee camp. They will target camp residents, as well as aid agencies, community leaders, block leaders, camp management, and other people in positions of power who are potential perpetrators. How will this help protect people, and what do you suggest they do?

2 You are working for an aid agency. After work you meet friends at a bar owned by the mayor's son. You notice many young women from other countries working in the bar. Some appear to be working as prostitutes, and others are dancers or serving drinks. You feel something is wrong, the women look frightened and unhappy and do not appear to be willingly working at the bar. The next day you raise the issue at an inter-agency SGBV working group. What kind of actions can the group take in this situation?

3 You work in a country recovering from a vicious and prolonged conflict in which there was widespread sexual violence. While progress is being made on political reform and rebuilding of the country, the attitude towards sexual violence has not changed. Although political leaders acknowledge that there is a problem, they are not addressing it as a priority issue. What can NGOs do to support national action to prevent, reduce, or respond to sexual violence?

4 You are working with displaced people living in camps in a conflict situation, and hear that women and girls are continuously subjected to rape and other sexual violence from armed perpetrators when collecting firewood outside the camp. They have to go further and further to search for fuel since there is little vegetation left around the camps. What can humanitarian organisations do to reduce the risk of sexual violence targeted at women and girls?

# Group 2: Displacement

This session is made up of small-group work based on a case study followed by a presentation in plenary.

## Trainer's notes

To deliver this session you will need:

1   to copy the case study on p. 131 for each group member

2   to draw the feedback template on p. 128 on a flip chart

3   to copy Handout 11 for each participant in the workshop

4   to draw the protection analysis template from p. 68 on a flip chart

5   to copy the programming options cards from p. 132

6   to provide flip-chart paper and pens.


Ideally there should be 4–8 people in this group sitting around a table or in a circle.


## Resources materials for trainers

▪ Handout 11: Supporting durable solutions to displacement

▪ The 'Reach Out Refugee Protection Training Kit' includes a module on durable solutions for refugees and an optional module on internally displaced people that supplement the materials included in this pack.

▪ The 'Internal Displacement Monitoring Centre' is the most comprehensive available source of information on internal displacement. Further training materials on IDPs and protection can be accessed from the website: www.internal-displacement.org

## Session plan

## Brief the group:

In crisis situations one way that people protect themselves and their families is by fleeing to another area (internally displaced people), or across an international border to seek asylum in another country (refugees). People can also become displaced if they are expelled from their homes, or coerced into moving to another area or country. Displaced people and refugees are particularly vulnerable because they often flee in situations of great danger, can become separated from their families and communities, have to abandon their assets and means of livelihood, and often do not retain documents that prove their legal status and identity. In some cases people can be displaced for many years, others are able to return to their homes if conditions allow it, or can integrate into the community to which they have fled, or resettle elsewhere.

This exercise will not look at ways of protecting people from forced displacement, but will focus on how humanitarian organisations can support durable solutions to displacement when it happens. In this exercise the group members will carry out a protection analysis of a case study and examine a range of programming options that support durable solutions for displaced people.

## The group work:

The facilitator should give out Handout 11 and talk through the handout with the participants. They will need to understand this information in order to complete the group work. The group members should now complete the template for feedback on durable solutions to displacement, which will form part of their feedback presentation.

Give out the case study and allow the group members enough time to read it through. The group members should then carry out a protection analysis using the template provided.

Once the protection analysis is complete, tell the participants that they are part of an inter-agency working group leading the humanitarian response to the situation. They have been asked to recommend actions to support durable solutions for the refugees.

Give out the programming options cards. Tell participants that they have enough collective resources to carry out five immediate actions. They should use the protection analysis to help them choose priority actions. They can also propose their own programming options in addition to cards they have been given. They should consider the assumptions being made for each activity (e.g. that people can read, or have radios).

In plenary the group should structure their feedback as follows:

1  A brief presentation on supporting durable solutions to displacement using the template (5 mins)
2  A summary of the situation described in the case study using the protection analysis format (5 mins)
3  An explanation of the five actions chosen and why they have chosen these options (10 mins)

During feedback, group members should be encouraged to be brief and to make their feedback interactive and lively. At the end of the presentation the group members should give the handout to all other participants in the workshop for future reference. Allow time for questions.

# Template for feedback

| Why do people leave their homes and where do they go? *(explain the difference between refugees and internally displaced people)* | What factors make displaced people and refugees particularly vulnerable? |
|---|---|
| | |
| What are the three main solutions for displacement and what conditions should be in place to ensure that the solution is durable? *(Refer to Handout 2, p.44)* | What factors limit people's ability to return to their homes? |
| | |

☞ **handout 11**

# Supporting durable solutions to displacement

One way that people protect themselves and their families is by fleeing either to another area, or across an international border to seek asylum in another country. Sometimes people are expelled from their homes, coerced or forced into moving to another area or country.

People who are displaced within the country in which they normally live are referred to as internally displaced people (IDPs). In most cases this is simply a description, not a legal status. **The UN's Guiding Principles on Internal Displacement** (1998) set standards for protecting people from displacement, the provision of assistance to and protection of IDPs during displacement, return and resettlement. They are not legally binding, but they do reflect existing international human rights and humanitarian standards.

People who flee across an international border to seek protection from another state are referred to as refugees. This is a legal description and offers special protection under the law. According to the **1951 Refugee Convention** the only grounds for conferring refugee status on a person is that they are fleeing persecution. However, the 1969 Organisation of African Unity Refugee Convention and 1984 Cartagena Declaration on Refugees extend this definition to include people fleeing internal and international conflict in Africa and Latin America. All three bodies of law set standards for the protection of refugees.

## Internally displaced people and refugees are particularly vulnerable because:

- they often have to flee in situations of great danger, through or within areas of active conflict, while under attack, and may have to go into hiding or move to other unsafe areas;
- they can face hostility from host communities including physical threats to their safety;
- displacement breaks up social and community leadership systems and protective mechanisms such as extended family networks, kinships groups, etc., although these may be rebuilt at a later stage;
- often people become separated from their family, friends, and other community members who may have provided protection for them during and after displacement. Women can be separated from their husbands, children from their parents, elderly people from their children. In the chaos of a sudden displacement some people (especially children) may become lost;
- when people flee they often have to abandon their assets and means of livelihoods (e.g. jobs, land, property, businesses) or these may be confiscated by parties to the conflict;
- during and following displacement people may lack access to basic services (food, water, sanitation) and facilities such as schools and clinics;
- people may have to abandon documents that determine their legal status and rights – identity cards and papers, passports, land and property ownership documents, etc.

Given the scale of displacement globally, the challenge of finding long-lasting solutions for displaced people is enormous. The main options for providing durable solutions to displacement are:

- **Voluntary return:** people voluntarily return to their homes when they feel it is safe to do so. This may be organised or spontaneous.
- **Integration:** people decide voluntarily to settle in the host community.
- **Resettlement:** including formal UNHCR-supported programmes for refugee resettlement in another state; formal programmes for the relocation of IDPs to a third location within a country; voluntary and spontaneous onwards movement to another location within a state.

Although the exact circumstances will change in each situation, some general principles must always be applied: all solutions must be voluntary – based on free and informed decision-making without coercion (including withdrawal of services or assistance); people must be assured of their physical safety, be free from persecution, and have access to their land or other means of meeting their basic needs; people must also be treated with dignity and respect and have their human rights respected and upheld, including clarity about legal status and citizenship. In all solutions the needs and priorities of especially vulnerable individuals such as elderly and disabled people, unaccompanied children, and ethnic minorities should be identified and addressed.

## NGOs can support durable solutions to displacement in many ways:

- By providing information: durable solutions are based on people having adequate, unbiased information about the options available to them in order that they can make free and informed decisions (including 'come and see' visits)
- By intervening with the responsible authorities when the conditions for return, resettlement, or integration are inadequate
- By providing assistance/services to prevent coerced/forced return or resettlement
- By providing assistance/services to support voluntary return or resettlement
- By providing assistance/services to support (re)integration and minimise tensions with host or remaining populations
- By supporting livelihoods options
- By helping people obtain legal status and documentation (including identity papers, land and property ownership papers, etc.)
- By supporting processes that enable people to regain their property, land, and homes or enable them to claim suitable compensation for losses

Although many people's preferred option is to return to their original homes, there are a number of factors that limit return, such as difficulties in regaining property or land; the occupation of homes by others; destruction of land, homes, and basic infrastructure (wells, schools, clinics, etc.); lack of access to employment or livelihood opportunities; physical insecurity, continual discrimination, or harassment and trauma.

## Resources:

The 'Reach Out Refugee Protection Training Kit' includes training materials on durable solutions for refugees, including the formal resettlement of refugees to third states, and has an optional module on internally displaced people.

The 'Internal Displacement Monitoring Centre' is the most comprehensive available source of information on internal displacement. Further materials on IDPs and protection can be accessed from the website: www.internal-displacement.org

# Case study: supporting durable solutions to displacement

Following an intense war, part of Country A seceded as an independent state known as Country B. This resulted in an exodus of 35,000 people who did not support independence to Country A, where they live in government-managed refugee camps along the new border. UNHCR has very limited access to the refugee population in the camps due to concerns about staff security. The government of Country A is seeking a long-term solution for the refugees and has set a time limit of six months for refugees to:

1  return to their original homes across the border, or
2  integrate within the region, or
3  resettle in other parts of Country A.

After the six months the government will end all support to the refugee camps, and they will be closed. Basic facilities and services in the camps are inadequate, but the government will not allow them to be upgraded as they believe this will encourage people to stay in camps.

Interviews with the refugees show that many are not aware of the deadline for camp closure and have little or no information about their options. Female-headed households, older and disabled people in particular have very little information about the situation and options available to them. The government has not provided any information to the refugees and has said that it will work through the self-appointed (male) camp leaders when it does so. Some camp leaders are former members of the militia who will be prosecuted for war crimes if they return to Country B. They have built considerable power bases in the refugee camps and are actively discouraging return, by means of misinformation, and circulating rumours of violent acts of retribution against refugees who go back to their original homes.

Some new homes are being constructed by the government to encourage resettlement, but many are on poor land in remote areas that lack basic infrastructure, and only a small number will be completed within the six months. Communities in the proposed resettlement sites are concerned about their resources (e.g. water, fertile land) being diverted to meet the needs of resettling refugees. UNHCR is funding the government's resettlement programme.

Informal discussions with refugees indicate that they would prefer to stay in the area around the camp and integrate into the local community because of cultural similarities and proximity to family members across the border. However, the local population itself is very poor, there is limited land available for integration, and there have been conflicts with the local community over resources, which have been severely stretched since the refugees arrived.

You have been asked to assess how national and international NGOs can support durable solutions for the refugees by:

- helping them to make informed decisions about their future and to act on those decisions;

- improving the capacity of the authorities to support return, integration, or resettlement;

- improving dialogue between governments, host/resettlement/return communities, and the refugees to ensure that all solutions respect the rights of the refugees and safeguard their dignity and safety.

# Programming options cards

| | | | |
|---|---|---|---|
| **Create international awareness to ensure that both countries uphold rights of refugees and returnees** | **Run an information campaign on refugees' options using radio, newsletters, etc.** | **Facilitate workshops with all stakeholders to look at how each option will be implemented** | **Monitor returns and resettlement to ensure safety, dignity, and respect for refugees' rights** |
| **Fund legal clinics to help refugees claim compensation for loss of assets in Country B** | **Monitor knowledge about each option and address identified gaps** | **Produce videos of 'come and see' visits and screen them in refugee camps** | **Fund a fleet of minibuses to transport people back to Country B** |
| **Act as a channel of information from government to refugees, including feedback on how each option is working** | **Survey refugees' preferences and lobby the government to resource each option** | **Run focus groups for women, older people, and disabled people to give information on their options and hear their concerns** | **Provide relief assistance in resettlement sites** |
| **Lobby the government to upgrade facilities in camps or allow others to do so** | **Organise 'come and see' visits to Country B and resettlement sites** | **Lobby the government of Country B to support returning refugees** | **Facilitate dialogue between host and resettlement communities on resource management** |

# Session 2: Objective setting

In this session, participants design their own programmes to respond to the threats that they identified in the protection analysis in Module 2, p. 68.

Using a basic logical framework they will identify protection objectives, outcomes, outputs, and activities. In Session 3 they will look at indicators and sources of verification and add these to the logical framework.

## Trainer's notes

To deliver this session you will need:

1  to redistribute the protection analysis templates that participants filled in during Module 2

2  to draw the logical framework template on a flip chart (one for each group)

3  copies of Handout 12 (p. 137).

## Resource materials for trainers

- Handout 12: Terms used in objective setting

- H. Slim and A. Bonwick (2005) *Protection: An ALNAP Guide for Humanitarian Agencies*, London: Overseas Development Institute, pp. 71–98.

## Session plan

The objective of all protection projects is that 'the safety of civilians is improved'. Objectives describe what the project wants to achieve over time (short or longer term) through collaboration and co-ordination of efforts by different actors.

Participants now decide on one or more outcomes of the project. Outcomes are what needs to happen so that people can lead safer, more dignified lives. The outcomes directly relate to the threat that is being addressed (e.g. gender-based violence, forced labour).

Participants should look at the protection analysis flip charts. The outcomes they decide on should reflect what the group aims to do together to achieve their objective in the area where they work, using their own expertise. They should refer directly to the threats and gaps they identified in the protection analysis in Module 2. For example: 'People in X are less at risk of forced labour in timber and rubber plantations' (don't worry about exact wording, so long as the meaning is clear).

Ask each group to think of three or four outputs. Outputs help to achieve the outcomes by reducing the level of threat, the time exposed to the threat, or the level of vulnerability. You may need to give some examples such as:

- Communities near to rubber and timber plantations understand their labour rights better
- The police force understands the risk of forced labour and what national law expects them to do about it
- UN peacekeepers are aware of the threat and impact of forced labour and actively address the risk to communities in action plans.

Once the outputs are agreed, each group should think of at least three actions that achieve the output in the short, medium, and longer term. Remind participants of the six different actions a humanitarian organisation may use:

- Advocacy
- Capacity-building
- Presence
- Assistance
- Voice
- Information

The facilitator should help the group tighten up their protection outcomes, outputs, and activities by asking questions like 'What about that is protective?' and 'How will that make people safer?'

# Never doubt that a small group of thoughtful, committed citizens can change the world; indeed, it's the only thing that ever has.

Margaret Mead

## Template for feedback

| Objective:<br><br>The safety of civilians is improved | | |
| --- | --- | --- |
| Outcomes: | Indicators: | Sources of verification: |
| Outputs: | | |
| Activities: | | |

☞ **handout 12**

# Terms used in objective setting

There are many different versions of logical frameworks (logframes). All follow the same general principles and stages, but they often use different terms that can cause some confusion. Here is a guide to the words you may come across and what they actually mean along with an example.

| Terms used | What it means | Examples |
|---|---|---|
| **Objective**<br>(sometimes called **Goal**<br>or **Overall Objective**) | What everyone involved hopes to achieve together over the longer term. Each actor will contribute towards this goal. | The safety of civilians is improved. |
| **Outcome**<br>(sometimes called **Specific Objective** or **Purpose**) | The change this project will achieve. | Humanitarian organisations have increased capacity to implement protection programmes. |
| **Outputs**<br>(sometimes called **Results**) | The different results of the activities. Each one contributes towards achieving the Outcome. | ▪ Humanitarian workers have increased knowledge of protection programming.<br><br>▪ Humanitarian workers have a range of tools and skills for implementing protection programmes.<br><br>▪ Humanitarian workers have effective national co-ordination mechanisms for planning, sharing skills and information and about human rights. |
| **Activities** | The practical things we do. | ▪ Carry out training workshops with humanitarian workers.<br><br>▪ Design and produce protection training packs.<br><br>▪ Establish national protection working groups. |

# Session 3: Indicators and monitoring

This session begins with a trainer-led discussion followed by small-group work.

## Trainer's notes

To deliver this session you will need:

1   the semi-complete templates for feedback on p. 136 used in previous sessions

2   copies of Handout 13 (p. 140)

## Resource materials for trainers

▨   Handout 13: Indicators and monitoring

▨   Useful resources on project cycle management are listed on Handout 6.

⧖ 90 mins

# Session plan

Start by summarising the work of previous sessions.

We have done a protection assessment and now know who is vulnerable and why, what threats they face, and their impact.

We have a clear objective for our programme, we have planned the outcome and outputs we want to see and have even suggested some activities we might carry out to achieve this.

But how will we know if we have been successful? How will we know if the activities we have decided on are working or if they need changing? What do we need to do if the external situation changes?

Introduce the idea of *monitoring*. Ensure you make the distinction between monitoring as a human-rights activity, and monitoring as part of the programme cycle. Ask participants to identify what would need to be done at the different stages of the programme cycle (assessment, design and resourcing, implementation, and end of programme) in order to understand whether or not a programme had achieved its protection objectives.

The groups should identify the following key activities:

- collection of baseline data
- design of indicators and sources of verification
- participatory planning sessions with stakeholders and key informants
- budgets and programme plans to include enough time for monitoring and cover the costs of human and other resources for monitoring
- establishing and using monitoring systems
- managers should use monitoring information to adapt and change protection programmes as needs and situations change
- evaluations or similar learning processes should take place.

In plenary take one or two outputs from the previous session and ask participants to suggest how we would know if we have achieved them. Introduce the idea of indicators, and sources of verification.

Participants then move back to their groups and decide on indicators for their outputs and outcomes (NOT activities). The facilitator should move between groups challenging the indicators and asking participants to think about how they would practically use them, and what sources of verification they would use.

Once the groups have decided on indicators they should feed back in plenary.

Ask other participants to challenge and make suggestions to strengthen their indicators.

☞ **handout 13**

# Indicators and monitoring

This handout is about monitoring protection programmes and projects. Monitoring allows us to measure progress so that we can adjust our projects to changes that are taking place. It also helps us to judge the impact of our projects and shows us how that impact was achieved.

There are various ways of measuring the impact of our programmes. Sometimes we must measure the numbers of abuses. But we must be careful with statistics. For example, the number of reported rapes may be much lower than the number of actual rapes. Often women don't want to report attacks for fear of what people might think about them, or because it can put them at further risk. A more useful way to measure the fear of rape may be to look at behaviour: Are women going out alone? Are there places they avoid? Or we can ask how many women are afraid that they or their family members might be raped.

Talking with groups of women can give useful information about what they fear and why. People can be afraid of things even if those things have not happened, and the fear of violence or rape, for example, can have a huge impact on people's lives. (It is important to remember that it is not only women or girls who are at risk of rape.)

In fact, protection programmes most often deal with abstract concepts such as 'fear of violence' or 'free and informed decision-making'. People experience these concepts very differently and they are hard to measure. So to monitor programmes addressing such issues properly, we need to use both *qualitative* (asking people what they feel, watching their behaviour etc.), and *quantitative* (counting numbers) 'indicators'. (There is more about this on p. 142.) It is best to use different sources of information and to check them against each other.

# What you need for monitoring:

1 **Baseline information**: what was the situation before the programme started? The programme may aim to make the situation better or just stop it from getting worse; either way you need a 'baseline' from which change can be measured. Usually baseline data is gathered at the beginning, when you are assessing what is happening.

2 **Resources**: it often takes much more time than we think to gather good quality information. Monitoring also needs people to do the work, who may need to be trained, and will certainly need to be managed and supported. It also requires support workers such as drivers, accountants to manage costs, and administrators.

3 **Management**: the information gathered during monitoring needs to be analysed and interpreted, and then it needs to be used. When you are planning programme monitoring you need to identify who will use the information and how. The person receiving the information needs to know how it has been interpreted, what assumptions have been made, and what are the limitations and weaknesses in the process, so that they can use it to make decisions particularly about adapting and changing the programme.

4 **Learning**: monitoring is also about learning. Changes in what is happening may mean a programme needs to be adapted. Also, learning while running a programme can show areas where doing things differently could make it better. Information gathered during monitoring can improve understanding of a situation and even show that some assumptions on which a programme was designed were wrong.

5 **Evaluation and impact assessment**: at the end of the project we need to understand what impact it has had, and what we can learn from the experience. Evaluation is often done by different people as part of a separate process, but the information gathered during monitoring will be an important record of the project and is crucial for the evaluation.

Indicators are used to monitor changes against the baseline, and to assess the impact of activities.

☞ **handout 13**

# There are two types of indicators

■ Qualitative indicators are useful to measure feeling, behaviour, and perceptions of the environment – such as freedom from fear

■ Quantitative indicators are useful to measure actions such as the number of people accessing a service, children attending school, or whether a policy has been changed.

Some examples of indicators for protection are outlined below:

| Example output or outcome | Types of indicators | Sources of verification for the indicator |
|---|---|---|
| **Free and informed decision-making** | **Qualitative indicators**<br>■ Access to impartial information<br>■ Ability to understand different choices<br>■ Confidence to make decisions without coercion<br>**Quantitative indicators**<br>■ Number of people receiving information<br>■ % of people understanding the choices<br>■ % of IDPs that act on information and make decisions | ■ Surveys<br>■ Focus-group discussions<br>■ Structured interviews<br>■ Analysing queries to see what information gaps exist. 'Informed' is easier to quantify than 'free', which is more subjective (different people mean different things when they say 'free'). |
| **Freedom from fear or threat of violence** | **Qualitative indicators**<br>■ People feel safe<br>■ People continue their usual daily activities (going to market, school, working in fields)<br>**Quantitative indicators**<br>■ Number of times a particular abuse or violation happens<br>■ % of children attending school, traders going to market, etc. It is important to disaggregate information by gender, age, and group.(e.g., maybe one particular ethnic group is fearful, while others behave as normal.) | ■ Focus-group discussions, surveys, interviews, ranking, scoring, pocket-voting<br>■ Studying behaviour through observation, or school attendance records, ethnicity of traders hiring market stalls, closing time of markets, etc.<br>■ Statistics from police/authorities, medical centres, etc. |
| **Safe access to services (water, medical clinics, etc.)** | **Qualitative indicators**<br>■ People feel safe using services<br>**Quantitative indicators**<br>■ Number of incidents of a particular abuse or violation while using services<br>■ % of people using the service<br>■ Sphere standards provide quality and quantity indicators but you need to define 'safe access' in each context and use additional indicators to specifically monitor it. | ■ Focus-group discussions, surveys, interviews, ranking, scoring, pocket-voting<br>■ Observation of service users<br>■ Statistics from police/authorities, medical centres, etc.<br>■ Statistics from service providers |
| **Changes in policy and practice of states and authorities, or perpetrators** | **Qualitative indicators**<br>■ Interpretation or understanding of policy is changed<br>**Quantitative indicators**<br>■ Policy is changed as desired<br>■ Changes in practice can be observed in advocacy. It may be possible to measure whether or not the desired change has taken place, but it is much harder to work out which action or organisation caused the change. | ■ Interviews, focus-group discussions<br>■ Studying and observing behaviour<br>■ Policy or legal documents and declarations<br>■ Sometimes it is possible to work out what or who caused the change by simply asking policy makers what influenced them. |

# Core exercises

This section contains four core exercises that reinforce and build on the learning in the previous modules. These exercises are not optional extras – they are central to the training programme. Each has been designed to achieve a specific goal:

- to set the tone and atmosphere of the workshop

- to challenge beliefs, values and attitudes about issues central to protection programming

- to reinforce key facts and draw attention to areas where there is disagreement about protection

- to help participants work out how to carry out protection actions in real-life situations.

# Using the core exercises

These exercises all reinforce learning points from the workshops and are designed to encourage interaction, and to create debate and discussion on some quite controversial and challenging issues. None of the exercises outlined here requires any physical contact between participants. Some require movement, but can be adapted for groups with less mobility. These exercises are central to the training, but have been designed also to act as icebreakers and energisers.

- **True or false?**: This exercise is used to reinforce key learning points and create discussion about contentious issues and misconceptions. It works well as an energiser after lunch.

- **Agree or disagree?**: This exercise is used to explore difficult or contentious issues, beliefs, and attitudes, and there are not necessarily right or wrong answers. The exercise can be used with a group that finishes ahead of others in group-work sessions, or it can be a stand-alone activity for all participants.

- **Protection stories**: These are practical examples of protection activities from Oxfam GB's work. They can be given out to participants to read or printed out as large posters and stuck up around the room.

There are two exercises which are described in full in earlier sessions, but they can also be used at other points during the training.

- **Inspirational quotes**: See page 19 for the full text for this session. This exercise creates a positive and motivating atmosphere.

- **What would you do?**: See page 108 for the full text for this session. This exercise reinforces key messages from the training.

# True or false?

This is an exercise that involves moving around the room, and can be used to energise a group while reinforcing key learning points.

## Trainer's notes

To deliver this session you will need:

1 to place the 'true', 'false', and 'don't know' cards in different parts of the room

2 to read through the 'Questions and additional notes' section on p. 150 to help you manage this session.

⧗ 20–60 mins

# Session plan

The facilitator reads out a series of statements. If participants believe the statement to be true they go and stand by the 'true' sign, if they believe that statement to be false they go and stand under the 'false' sign. The facilitator then picks one person from each group to explain why they believe the statement to be true or false.

A third 'don't know' sign can be used. Participants who don't know whether a statement is true or false stand by this sign (place between the others) and participants from each of the 'true' and 'false' groups try to persuade them to join their group. (This is optional and requires more time for discussion.)

There are two sets of questions relating to:

1 international and national law and other standards for the protection of civilians

2 the roles and responsibilities of protection actors.

There are notes to help the trainer answer participants' questions, but it will still be necessary for one of the trainers to have a good knowledge of the relevant bodies of law and mandates of agencies such as UNHCR in order to carry out this exercise, or for an additional resource person with this knowledge to take part in these sessions. Some statements have a clear 'true' or 'false' answer, but others are less clear depending on the exact situation or the interpretation of the law.

This excercise can be adapted for small rooms or participants with limited mobility. Give each participant a card with 'true' on one side and 'false' on the other. Ask them to hold up the relevant answer to each statement. Those who 'don't know' don't hold up a sign.

# True or false statements

## Questions and additional notes

### Set 1: International standards for civilian protection

**1 Displaced people don't have the same rights as refugees**

**Answer: False/True.** Human-rights law applies to all people regardless of whether they are refugees or displaced. However, 'refugee' is a legal status and the rights of refugees and protection that should be granted to them are determined by the 1951 Refugee Convention (and the Cartagena Declaration for refugees in Latin America, or OAU Refugee Convention for refugees in Africa). The term 'displaced person' is a description, not a legal status, and therefore does not grant displaced people specific legal rights. The UN has developed Guiding Principles on Internal Displacement, which are not legally binding (unless a state brings them into national law as in the case of Angola) but are authoritative and reinforce existing protection in human rights and international humanitarian law.

**2 It is not illegal to kill civilians in war**

**Answer: True.** International humanitarian law (IHL) (notably the Geneva Conventions and Additional Protocols) was created specifically to protect civilians and other people not fighting or no longer fighting in times of war. IHL requires parties to the conflict never to deliberately attack civilians or civilian assets, and even when attacking military objects or assets they must take all reasonable precaution to avoid hurting civilians. Therefore it is not necessarily illegal to kill civilians in war, but the deliberate targeting of civilians is illegal and constitutes a war crime.

**3 More than 1,000 people have to be killed for genocide to have taken place**

**Answer: False.** Genocide is defined as committing acts with intent to destroy, in whole or in part, a national, ethnic, racial, or religious group. The acts that construe genocide are: killing, causing serious bodily or mental harm, deliberately inflicting conditions of life calculated to bring about physical destruction in whole or in part, imposing measures to prevent births, or forcibly transferring children from the national, ethnic, racial, or religious group being targeted. Therefore intent is the key determinant of genocide – not numbers of people affected.

**4 States are able to use torture in exceptional circumstances such as a national emergency**

**Answer: False.** The Torture Convention bans the use of torture in all circumstances. Human rights, including the right not to be subjected to cruel, inhuman, or degrading treatment or punishment, apply in both peace time and war, although it is possible for a state to derogate some rights in cases of national emergency that threaten the life of the nation. There are some human rights which can never be derogated called peremptory norms or non-derogable rights – they include the right to life, freedom from torture and slavery, recognition as a person before the law, and freedom of thought, conscience, and religion.

**5 The International Criminal Court will only be able to try a few war criminals**

**Answer: True.** The Prosecutor of the International Criminal Court (ICC) has the discretion to investigate war crimes, crimes against humanity, and genocide in states that are party to the statute of the court where that state is unwilling or unable to act itself. The UN Security Council can also request the Prosecutor to investigate crimes elsewhere, as is the case in Darfur, Sudan. However, due to its limited resources, the ICC is likely to focus on prosecuting the most serious crimes, and the most prominent, serious, or high-ranking offenders.

**6 African women don't have a specific law to protect them against discrimination**

**Answer: True/False.** While there is no Africa-specific version of the Convention on the Elimination of all forms of Discrimination Against Women (CEDAW), CEDAW is an international law with global applicability. In addition, the African Charter on Human and Peoples' Rights, Article 18(3), states: 'The state shall ensure the elimination of every discrimination against women and also ensure the protection of the rights of the woman and the child as stipulated in international declarations and conventions.'

**7 Sexual violence, such as rape, is always a violation of international law**

**Answer: True.** The prohibition of sexual violence is found in several international instruments, including the Geneva Conventions and the Convention on the Rights of the Child (Arts. 19, 34). The prohibition on sexual violence falls under the well-established prohibition on torture and cruel, inhuman, and degrading treatment. The prohibition is so widespread that it is considered part of international customary law, which means that it applies even to countries that have not yet endorsed it. In addition, the International Criminal Court recognises sexual violence as a war crime and a crime against humanity in certain circumstances. Rape is also a crime in most countries' national law.

**8 In war, threatening a civilian with violence is against international humanitarian law**

**Answer: True.** The Geneva Conventions and the Additional Protocols strictly prohibit violence, murder, and torture – whether physical or mental, mutilation, humiliating and degrading treatment, enforced prostitution, rape, taking of hostages, collective punishments, as well as threats to commit any of these acts.

**9 A woman who has lived in a displacement camp for five years is no longer an IDP (internally displaced person)**

**Answer: No answer.** There is no time limit on how long you can be an IDP. The Guiding Principles don't specify how long someone can be an IDP, nor what makes them no longer an IDP. IDP is a description and not a legal status (unlike 'refugee' which is an internationally agreed legal status), therefore it is difficult to determine officially when someone stops being an IDP. In most cases people stop being described as IDPs when they permanently resettle in another location or return to their original homes because the factors that caused their displacement have ended (e.g. end of a conflict).

**10 There is no way to monitor whether a government abides by the Convention on the Rights of the Child (CRC)**

**Answer: False.** All states that have become parties to the CRC are required to submit reports every five years to the Committee on the Rights of the Child. These reports are available on the Internet for anyone to read on the UNHCR website: www.unhchr.ch/tbs/doc.nsf. NGOs can provide information to the Committee on the Rights of the Child and this information is taken into account when the Committee responds to the state's original report with its recommendations for action in order to fulfil obligations as a signatory to the CRC.

**11 A state can launch military action against another country that is violating the human rights of its citizens**

**Answer: True/False.** The UN Charter bans the use of force between states except in self-defence or when authorised by the Security Council. Therefore a state can only launch military action against another state if the Security Council has authorised it to do so. However, at the UN World Summit in 2005, states agreed that the Security Council should authorise the use of force when faced with widespread human-rights violations. Criteria for this are set down in a series of documents entitled Responsibility to Protect (see www.responsibilitytoprotect.org for further information).

**12 Even if a country already has more refugees than it can cope with, it cannot close its border to new asylum seekers**

**Answer: True.** Everyone has the right to seek asylum from persecution (Universal Declaration of Human Rights Art. 14). The principle of non-refoulement is central to refugee law – states cannot return people to a place where they will be at risk of persecution.

**13 A national army cannot conscript 17-year-olds into its armed forces for active service in conflict**

**Answer: Depends.** International humanitarian law sets the minimum age for participating in hostilities at 15 years and prohibits the recruitment into the armed forces, and the direct or indirect participation of children under 15 in hostilities. Both IHL and the Convention on the Rights of the Child require states to give priority in recruitment among children aged 15 to 18 to the oldest. However, if the state in question is party to the Optional Protocol to the Convention on the Rights of the Child on the Involvement of Children in Armed Conflict (2000) it must take all feasible measures to ensure that members of the armed forces who have not reached 18 do not take direct part in the hostilities; that compulsory recruitment into the armed forces of children under 18 is prohibited; and that the minimum age for voluntary recruitment is 15 years (does not apply to military academies). Non-state armed forces should not recruit (compulsorily or voluntarily) any person under 18 years of age in any circumstance. Eighty-eight states have ratified the Optional Protocol on Children in Armed Conflict including Uganda and the Democratic Republic of Congo. A number of other states have signed, but not yet ratified, the Optional Protocol.

The Statute of the International Criminal Court includes in its list of war crimes within the Court's jurisdiction the active involvement in hostilities of children under 15 or their recruitment into national armed forces or other armed groups during armed conflict.

# Set 2: Roles and responsibilities among protection actors

**1   UNHCR is legally responsible for making sure that refugee girls aren't abducted by rebel groups**
**Answer: False.** Legal responsibility for the protection of refugees and IDPs – including keeping armed groups out of camps – always lies with the host country, although UNHCR, ICRC, NGOs, or neutral states can help provide protection assistance.

**2   The government is responsible for keeping rebel groups outside of refugee camps**
**Answer: True.** Legal responsibility for the protection of refugees and IDPs – including keeping armed groups out of camps – always lies with the host country. UNHCR holds a mandate – but not legal responsibility – to provide protection; and ICRC, NGOs, or neutral states can also help. The Protocol of the Convention on the Rights of the Child (Art. 4) as well as the African Charter on the Rights and Welfare of the Child (Art. 22) require states to take all necessary measures to prevent recruitment of children, including children in refugee camps.

**3   UN peacekeepers do not always have to protect civilians**
**Answer: True.** Most UN peacekeeping missions are now mandated to protect civilians under immediate threat of physical violence if it is within their capacity to do so. However, many peacekeeping missions are under-resourced and therefore don't have the capacity to protect civilians, e.g. in Rwanda in 1994 or in Srebrenica in Bosnia in 1995. Humanitarian organisations can play an important role in advocating for appropriate mandates and resources to protect civilians.

**4   International humanitarian organisations have the primary duty to provide humanitarian assistance to IDPs**
**Answer: False.** National authorities have primary duty and responsibility for this, under their own sovereignty. The UN acts in support, typically once a memorandum of understanding with the state has been signed. Many NGOs also take this approach although some (e.g. Médecins sans Frontières) maintain that they have a right or duty to intervene independent of the state.

**5   Rebel groups have to protect civilians**
**Answer: True.** Non-state armed actors may not be signatories to international humanitarian law, but they are still obliged to respect these laws, particularly with regard to protecting civilians, and the use of child soldiers. If they do not respect the law they can be prosecuted. For example, Charles Taylor has been indicted at the Special Criminal Court for Sierra Leone for acts committed when he was head of a rebel group.

## 6  Resolutions by the UN Security Council cannot be enforced because there is no global police force

**Answer: No answer.** States are obliged to respect UN Security Council resolutions; however, there are many cases where states are unwilling to do so. The UN can put pressure on states to fulfil their obligations through quiet diplomacy or coercive actions such as sanctions. Civil society can also put pressure on states to fulfil their obligations. Although the UN Charter envisaged the creation of a standing UN military force, this has never actually been created.

## 7  Humanitarian organisations have the legal right to deliver humanitarian assistance

**Answer: False.** Civilians have the legal right to access humanitarian assistance – the right belongs to the civilian not the humanitarian organisation. It is the state that has the primary responsibility for providing humanitarian assistance to its civilians. If the state is unwilling or unable, UN agencies, the Red Cross, and NGOs can deliver humanitarian assistance, but do so with the state's permission (which will not normally be refused).

## 8  The police have to investigate crimes in wartime as well as in peacetime

**Answer: True/Depends.** In order to protect people's human rights, the state should investigate any criminal activities or breaches of the law in the country. This is normally done by the police force, although in situations of martial law the army steps in. Crimes committed by members of the armed forces, including war crimes, are frequently investigated by specialist military police and prosecuted through courts martial.

## 9  It is the responsibility of the UN to collect hidden weapons in former war zones

**Answer: False.** It is the responsibility of the state to ensure that its citizens are safe, including controlling the availability and use of small arms. However, governments in war zones may not be able or willing to do this alone, and the UN and other international bodies (such as the European Union and the African Union) may provide support in disarmament processes, and in the creation of a safe environment.

## 10 Humanitarian organisations must provide humanitarian assistance to both sides in a conflict

**Answer: False.** The humanitarian imperative obliges humanitarian organisations to provide assistance wherever it is needed if it is within their ability to do so. Humanitarian assistance is not provided as a political or partisan act and is driven by the principles of humanity, independence, and impartiality. Impartiality means providing assistance based on people's needs and their capacity to meet those needs (as opposed to simply providing aid to all sides in a conflict).

# Agree or disagree?

This is a flexible energiser that can be used at any point in the workshop. It creates discussion and debate on issues in protection.

## Trainer's notes

To deliver this session you will need:

1 Agree or disagree cards (provided on pp. 162–5 and available in colour at the back of the book)

2 to read through the additional notes on pp. 157–61 beforehand to help you facilitate the session and highlight the key points for each statement.

⧖ 20–60 mins

# Session plan

This exercise can be used as a stand-alone exercise in small groups or pairs or in plenary, or can be used as a supplement to other exercises, for example, for participants who finish group work before others.

Ask participants to pick a card (or choose which cards go to which groups if you want to raise specific issues). The participants then discuss the statement on the card and decide whether they agree or disagree with it before feeding back to the whole group in plenary. Some statements will generate more discussion than others. If some participants finish discussing their statement sooner than others, they can just pick another statement.

**The trainer will need to guide participants as to how much discussion is required given the time available:**

To use the exercise quickly, give one statement to each pair or small group, and ask them to quickly discuss their initial reaction and feed back with one or two key points.

If you have more time available give each pair or small group a statement to discuss in detail and ask them to come up with two to three points in agreement and two to three points in disagreement with a statement.

Some of the statements deal with sensitive issues such as rape. If a participant is uncomfortable with the statement they have been given, perhaps because it relates very closely to their own experiences, allow them to exchange it. It may be useful to have a couple of quite neutral statements available for this purpose.

# Agree or disagree statements – additional notes

## People have fewer rights in war

International humanitarian law (most notably the Geneva Conventions) applies only in war or armed conflict. Human rights apply in both peacetime and war, although it is possible for a state to derogate (limit or stop) some rights in cases of national emergency that threaten the life of the nation. There are some human rights which can never be derogated, called 'peremptory norms' or 'non-derogable' rights – they include the right to life, freedom from cruel, inhuman, or degrading treatment or punishment and slavery, recognition as a person before the law, and freedom of thought, conscience, and religion.

## Domestic violence is a private matter within a family

Domestic violence is a form of gender-based violence and is a serious, life-threatening protection issue primarily affecting women and children. Gender-based violence is a violation of universal human rights (including the right to security of the person, the right to the highest attainable standard of physical and mental health, the right to freedom from torture or cruel, inhuman, or degrading treatment, and the right to life). Domestic violence is often more violent and more frequent after a disaster or in conflict situations. Domestic violence is also a criminal offence in most countries.

## All development and humanitarian work is ultimately protective

In the broadest sense it could be argued that all development and humanitarian work is ultimately protective in that it aims to help people attain their human rights. However, to argue this would risk losing sight of the need to protect people from real and immediate threats of violence, coercion, and deliberate deprivation. Everything we do affects the protection of civilians, but that does not mean that everything we do is protective.

## Protection is only done by specialists like UNHCR and ICRC

Both the UNHCR and the ICRC have mandates to protect civilians (in UNHCR's case specifically refugees, and recently it has also taken on responsibility for IDPs). However, NGOs and other actors can play an important role in keeping people safer – especially if all actors work together to complement each other's work and co-ordinate their efforts.

## The perpetrators of violence are usually men

Violence is often carried out by the most powerful in a society upon the most vulnerable. Gender roles in many cultures make women more vulnerable, as does women's position as child-bearers, and their physiology. Those in power are often men. Therefore we can see that those being violent are more often men, and acts of violence are more often aimed at (or, the victims are more often) the vulnerable in society – women, children, elderly people, disabled people, and minority groups. Pre-existing patterns of discrimination and vulnerability often get worse in times of conflict, as does gender stereotyping that encourages men to demonstrate physical strength as fighters and soldiers. While the violent may more often be men than women, not all men are violent and it is dangerous to assume that they are – in many conflicts women and children have also been violent, and men are frequently victims of violence. Men can also play an important role in protecting their families and communities or other communities, for example as peacekeepers.

## You have to be a legal expert to work on protection

When abuses take place against civilians it is often clearly wrong – for example the sexual slavery of girls and women, or the killing and wounding of civilians. You do not need legal knowledge to understand when abuses are taking place nor to analyse the situation and respond. However, it can be useful to have a basic knowledge of the relevant bodies of law as they form the framework for protection work. It also helps to work with protection core groups or working groups that can provide support and expertise about the law as well as other aspects of protection.

**You can use the law in three ways:**
- to set standards
- to locate responsibility
- as an argument of persuasion and tool for advocacy.

Protection working groups can help organisations working on protection to access appropriate legal expertise.

## It is the government's job to protect its citizens

It is primarily the responsibility of states to protect their citizens. Where states are unable or unwilling to do so, other states and state bodies (e.g. UN, EU, AU, ECOWAS) can offer support or can intervene to protect civilians.

## If you carry a weapon you are no longer a civilian

A civilian can be defined as a person not taking an active part in hostilities. Carrying a weapon does not necessarily mean that a person is taking an active part in hostilities – they could be a farmer carrying a gun to shoot animals, or a householder keeping a gun to protect his/her family or property. By taking active part in hostilities a person loses their civilian status.

### It is inevitable that, in times of war, some civilians will get hurt

International humanitarian law (notably the Geneva Conventions and additional Protocols) was created specifically to protect civilians and other non-combatants in times of war. IHL requires parties to the conflict never to deliberately attack civilians or civilian assets. Even when attacking military objects or assets they must take all reasonable precaution to avoid hurting civilians.

### Women and children are most at risk, therefore we should focus on protecting them first

In most cultures it is women and children who are among the most vulnerable members of society, but it is important to analyse every situation to establish which members of society are most vulnerable and why. We often find that children are most vulnerable because of their limited ability to protect themselves and their lack of experience of life. Women have specific vulnerabilities relating to their gender and particularly to their reproductive role. However, some violations may be specifically targeted at men (such as forced labour, forced conscription, and sexual violence) or they may be at risk because of a disability, or their ethnicity. It is important to carry out a detailed analysis of vulnerability and the way in which threats are targeted in order to identify those most at risk and support them appropriately.

### The world has a responsibility to help protect us

The 191 United Nations member states have a collective responsibility to maintain international peace and security. In recent years the UN has recognised that human-rights abuses within a state are a threat to international peace and security. For example, in the case of Liberia the UN Security Council acknowledged human-rights abuses as a threat to peace and security in the region. Where a state is unable or unwilling to protect its citizens, UN member states have agreed a collective responsibility for upholding human rights either by supporting the state or by other forms of intervention (e.g. peace-enforcement). It is increasingly argued that UN member states have a responsibility to protect, including the possibility of the use of force.

### People mainly protect themselves

Most people in armed conflict don't see outside actors or interventions and are not protected by them. Individuals, communities, families, kinship groups, tribes, and clans all play an important role in protecting civilians (though they can also be a source of threat). Civilians protect themselves from threats in three ways: by avoiding them, submitting to them, or confronting them. To make civilians safer, NGOs need to help them increase their own ability to protect themselves.

### Relief distributions can be used to protect people

Relief goods can be used to protect people both directly and indirectly. Directly, relief assistance can offer protection by:
- meeting basic human needs, which are also human rights – water, shelter, food, etc.
- preventing people from having to go into dangerous areas to meet their basic needs
- reducing vulnerability and exposure to threats.

Indirectly, providing assistance means NGOs can get access to populations at risk, so can monitor human rights in a specific location.

## Relief assistance can expose people to further violence

Relief assistance can protect people's rights (e.g. by providing food or shelter) but it can also expose people to further risk, for example, to violence during looting of the relief goods. Relief assistance can be given in a way that reduces risk, but because it can also increase people's risk, communities have sometimes rejected humanitarian assistance when they felt it made them more vulnerable to attack. Many people who are insecure put their own safety before assistance – assistance can put them at more risk unless security is guaranteed. Relief assistance can also expose people to exploitation, particularly sexual exploitation, by aid workers and members of the community controlling or involved in distributions.

## Protection is all about advocacy

Protection is about making people safer, and advocacy is just one of several ways to reduce the level of threat. Advocacy is best used in combination with other activities such as conscious presence and capacity-building, or activities to reduce vulnerability such as assistance, information sharing, and enabling communities at risk to speak out.

Protection is not just about what you do, but also about how you do it – a protection analysis puts the safety of civilians first and can result in a different type of intervention; for example, it may show that distributing relief items to people living in insecure conditions is inappropriate if it would increase the risk of violent attacks by those looting relief items.

## The best way to protect people is to give them guns

People protect themselves in different ways, by:

- avoiding a threat (e.g. fleeing or seeking asylum, by not tending their fields or travelling to market)
- submitting to a threat (e.g. joining a militia, paying taxes, handing over their sons, etc.)
- confronting a threat (e.g. in a non-violent way through protests, campaigns and similar action, or by physically protecting oneself and one's family with arms).

NGOs support some of these self-protective actions, but would never provide arms. In cases where arms have been provided to people at risk, it has simply created another threat to vulnerable civilians (e.g. northern Uganda). It is important to understand how removing weapons from groups can change the power balance and increase the vulnerability of certain groups; for example, during disarmament processes.

## If you want to find out whether people have been raped you should just ask them

Firstly, you need to be clear why you need to know whether people have been raped. It may be more appropriate to ask communities to tell you about abuses they are suffering, or fear, and to see if they voluntarily talk about rape or use other words to refer to rape or sexual assault. If you do have a good reason for asking whether people (male as well as female) have been raped, approach the issue with sensitivity and care.

The fear (or threat) of sexual violence can have as great an impact as actual incidents.

Sexual violence is vastly under-reported due to the shame and self-blame that many victims/survivors feel, their mistrust of authorities, and the fear of reprisals or further victimisation. It is therefore very difficult to obtain information about sexual violence happening. Recent guidelines for humanitarian actors working with gender-based violence recommend that they should assume that sexual violence is taking place and is a serious and life-threatening issue, whether or not there is reliable evidence.

Anyone collecting information about sexual violence should have support and training to do so, and be sensitive in doing interviews and collecting information – even giving information can put victims at further risk, and can be a difficult and traumatic experience.

One way of understanding whether a particular violation is taking place is to look at proxy indicators, such as whether women are changing their normal routines to avoid threats (not working in the fields or travelling to market for example), and information from secondary sources, such as medical NGOs who may be treating victims of sexual violence.

## 'Protection' is just a new term for what we've already been doing – it's just the latest trend

Humanitarian organisations have for years carried out much assistance and advocacy work that has helped protect people in armed conflict and post-conflict situations. However, analysing a situation with the specific aim of making people safer can result in a different set of activities carried out in a different manner, or in the inclusion of protection-specific activities. Through good analysis and systematic planning, humanitarian organisations can help make civilians safer.

# Agree or disagree cards

**People have fewer rights in war**

**Domestic violence is a private matter within a family**

**All development and humanitarian work is ultimately protective**

**Protection is only done by specialists like UNHCR and ICRC**

**The perpetrators of violence are usually men**

**You have to be a legal expert to work on protection**

**It is the government's job to protect its citizens**

**If you carry a weapon you are no longer a civilian**

| | |
|---|---|
| **It is inevitable that, in times of war, some civilians will get hurt** | **Women and children are most at risk, therefore we should focus on protecting them first** |
| **The world has a responsibility to help protect us** | **People mainly protect themselves** |

| | |
|---|---|
| **Relief distributions can be used to protect people** | **Relief assistance can expose people to further violence** |
| **Protection is all about advocacy** | **The best way to protect people is to give them guns** |
| **If you want to find out whether people have been raped you should just ask them** | **'Protection' is just a new term for what we've already been doing – it's just the latest trend** |

# Protection stories

This is a flexible resource that gives practical examples of protection activities from Oxfam GB's work in four countries. It can be used by the trainer at any point in the training to illustrate what protection means in practice. The examples can be printed out as posters and placed around the room for participants to refer to throughout the training. Alternatively the trainer can choose to give out copies of the case studies for participants to read and present as part of a structured session.

## Resource materials for trainers

Inter-agency Standing Committee (2002) 'Growing the Sheltering Tree: Protecting Rights Through Humanitarian Action', programmes and practices gathered from the field.

# Indonesia:

## debunking assumptions about refugees' needs in West Timor

In 2002, in West Timor, Indonesia, refugees displaced from East Timor in 1999 were under increasing pressure (for example, through the suspension of food aid) to disband camps and either resettle elsewhere in Indonesia or repatriate.

At the same time, the assisted repatriation scheme was stopped, as it was assumed that no more people would take it up. A new scheme was promoted to resettle refugees in other parts of Indonesia, away from West Timor. There was little knowledge of refugees' opinions and concerns, including whether they would accept an option to move off the island. There was very little information available on the readiness of the resettlement sites to receive the refugees.

The fact that refugees did not move, despite mounting pressure, indicated the degree of uncertainty refugees associated with leaving the camps. Oxfam was increasingly concerned that as time passed, more overt coercive measures would be used to disband the camps. Oxfam and a group of local NGO partners initiated a survey that helped to 'debunk' some assumptions held by stakeholders. It provided information about refugees' real needs, their priorities when considering local settlement or repatriation, and how decisions were reached in communities on these matters. Among other issues, the survey results highlighted the fact that refugees said that they would never consider resettling away from Timor island.

Given the vast shortage of reliable information, this effort to systematically collect information on refugees' needs and concerns produced significant changes in the government and humanitarian agencies' beliefs, ideas, and policies regarding the remaining refugees. In particular, the push to get refugees to move to other islands was all but dropped, assistance for repatriation continued, and local settlement in West Timor was finally acknowledged as an option that needed to be developed.

At a local level, dissemination activities about successful examples of spontaneous integration were shared through radio-dialogues, newsletters, and project teams, both to motivate refugees to take initiatives and also to motivate the government to support these initiatives.

This was complemented through increased activities to engage with local officials, for example, by producing and distributing an information sheet about the project and by carrying out more regular visits with village and sub-district leaders to share issues and update them on the situation of the refugees. These approaches have had positive effects: for example, two village chiefs offered land for local settlement in their villages.

# Sudan:
## reducing women's exposure to violence at water points

More than 40,000 IDPs fled to Kebkabiya town to seek safety from the violence engulfing the Darfur region of Sudan. There were some water points throughout town, however an Oxfam assessment found that the quantity and quality of water was insufficient. In addition, at most water points women and girls reported violence and harassment from the militia. The abuse included beating and whipping, water containers being confiscated, and shooting into the air to scare and intimidate. Women and young girls in and around the town were regularly abducted and gang raped for days at a time. Levels of violence were highest in isolated locations on the town periphery, where the nearest house is over 500m away, and near areas of high militia presence. Men would not leave the town boundary because of a very real threat of being shot.

As a result of this assessment, it was decided that Oxfam would not refurbish any water source that IDPs considered insecure. During the assessment of water points, many potential sources were eliminated from consideration due to their isolation and/or proximity to areas of high militia presence. Four viable water sources were ultimately identified which explicitly met the criteria of enabling safe access to water. In one case, Oxfam ran a pipeline from the town's outlying well into a residential area, in order to ensure safe access to water by IDPs.

Apart from standard water point monitoring, the project planned to monitor the extent to which women felt safer and less subject to harassment and violence at the new water points. This project aimed to benefit more than 4,000 households (24,500 people) and significantly reduce exposure to violence for women and young girls.

# Israel and the occupied Palestinian Territories:

## restricted movement in the West Bank

Oxfam was rehabilitating springs, increasing the water storage capacity, and supporting a women's 'water user group' to identify and address public health risks in Deir Sharaf, a Palestinian village near Nablus in the West Bank. The women identified a huge pile of rubbish dumped at the edge of their village as a major health hazard. The Israeli soldiers controlling the checkpoint between the village and the designated landfill site near a neighbouring village, had repeatedly prevented the rubbish trucks from passing through, despite attempts by the village council to liaise with the Israeli coordination office (DCO).

In conjunction with the village council, Oxfam paid a contractor to remove the rubbish to the landfill site, sending the relevant identity, registration and insurance numbers of all the drivers to the DCO to obtain their permission to pass the checkpoint.

Oxfam believes that until there is a political solution to the conflict in the occupied Palestinian Territories, long-term sustainable development will not be possible. Ideally, there should be no checkpoint prohibiting the movement of Palestinians to dispose of their rubbish.

Oxfam has therefore prioritised the protection of civilians' rights – as enshrined in international humanitarian law – as the most urgent need of the current humanitarian situation in the occupied Palestinian Territories. Moreover, Oxfam believes that the protection of civilians is not only their right, but is essential to build peace and reverse the humanitarian situation. Oxfam International published a policy paper on the issue (Briefing Paper no. 62) and is now advocating with Israeli and Palestinian civil society, as well as the UK Foreign and Commonwealth Office, and European Union members to urge parties to the conflict to put protection first, even if it means using a third party.

# Democratic Republic of Congo:
## the UN Security Council takes action

Ethnic clashes in Ituri, north-eastern DRC, have killed at least 50,000 and displaced 500,000 others since 1999. A vast humanitarian crisis coupled with difficulty of access has meant that Oxfam's public-health assistance programme could only scratch the surface of needs, and of course could do nothing to address the principal issues: protection of civilians from indiscriminate violence. Aware that needs were not limited to food, shelter, or water, Oxfam invested substantial resources over three years to become an authoritative source of information and a convincing advocate on the crisis.

In 2001, a timely intervention through a British ambassador with whom Oxfam had built a trusting relationship resulted in troops bringing a massacre to an end quickly. In 2002, Oxfam worked with the French ambassador to the United Nations and other diplomats to propose a resolution on DRC to the Security Council. Resolution UNSC 1445 was passed, and included much of the language that Oxfam proposed on Ituri, such as naming the parties to the conflict, the safety of humanitarian workers, access to those in need, and an increased presence in Ituri of MONUC (the UN Mission in the Democratic Republic of Congo). As the violence continued, in 2003, Oxfam supported the UN Department of Peacekeeping Operations in lobbying for a peace-enforcement response. This resulted in the first deployment of the European Union's Rapid Reaction Force: Operation Artemis. This force brought limited stability to the area, allowing humanitarian workers access to the main town, Bunia, and laying the political foundations for the installation of a two-year transitional national government.

Although the situation in DRC in general, and Ituri in particular, remains critical, Oxfam's intervention has had considerable life-saving impact, beyond its relief delivery in the fields of water and sanitation.

# Peace cannot exist without equality.

Edward Said

# Closing the workshop

The closure of the workshop is as important as the opening. This is when the trainer will go back to the agenda and summarise what has happened during the workshop, the key learning points, and what participants should be able to do with the skills they have learnt.

It is also an opportunity for the trainer to make some personal reflections on the workshop.

You may also find that some participants want to thank the trainers and may wish to make a formal statement to the group as representatives of their organisations or networks.

# Dealing with outstanding issues

As well as summarising what has been done, the trainer should summarise what has not been covered – any issues that were raised but 'parked' or left to be followed up outside the workshop. Ensure that you go through all these outstanding issues and agree what follow-up will take place, by whom, and when.

# Participants' contact details

You will have a list of contact details of participants from their registration, and may also have collected a list during the workshop. If it is possible to photocopy the list, do so and distribute to the participants before they leave. Encourage participants to exchange contact details directly as well.

# Feedback and evaluation

There are many different methods for obtaining feedback and most facilitators will have several methods they use. The reasons for obtaining feedback are to:

- understand how effective the workshop has been in the short term
- understand how enjoyable and informative it was for the participants
- find out how to make it better next time.

It is important to use feedback on the learning experience in order to improve future workshops. An example feedback form is provided on p. 176, which you can adapt if necessary. You can also use other methods. Here are some examples you can use or adapt:

- Prepare three flip charts with traffic lights and ask participants what you should stop (red), keep the same (amber), and change (green).

- Ask participants to discuss in pairs one thing they have learnt or will do differently as a result of the workshop. After five minutes of discussion ask each person to present in plenary either their own point or that of their partner.

- Remind participants about the inspiring quotes exercise at the start of the workshop. Ask them now to each say one thing from the workshop that will continue to inspire them in their work.

- Take four flip charts and label them 'Learnt most', 'Learnt least', 'Most fun thing', and 'Most boring thing', and ask them to choose one activity, session, or event to put on each flip chart.

- Write up statements on the wall such as 'I now understand what protection is', 'I know how to do a protection analysis', 'I can think of at least one thing I am going to do to improve the safety of civilians', etc., and ask participants to tick any statement they agree with.

As participants leave, stand near the door and say goodbye to each person individually, being sure to thank them for their particular contribution to the workshop.

# Practicalities

Once participants have gone home, you will be left in an empty room with flip charts stuck all over the wall, blu-tak and posters to sort out, and your training cards scattered around. Ideally you will have gathered up your training resources after each activity or exercise, checking that you have everything as you pack. In order to ensure that the training pack stays in good condition you'll need to:

■ remove posters from the wall carefully, peel off the blu-tak, and store them safely

■ count the activity cards after each workshop and hold them together with elastic bands.

If you lose any materials you can photocopy the master versions in the manual. The artwork is also on the CD for printing replacements. If you don't have access to a photocopier or printing facilities, you can simply write out new cards by hand.

# Trainers' feedback

After the workshop the trainers should give each other feedback on their performance. The format 'what should I continue doing, stop doing, start doing?' is useful as it de-personalises the feedback process and forces each trainer both to give and accept negative and positive feedback. As a result of examining the participants' feedback and trainers' feedback to each other, key points for improving future workshops should be recorded and shared among all trainers using these materials.

# Follow-up

If they have email, contact participants within a week of the workshop to thank them for their participation. This is also an opportunity to follow up on feedback and suggestions for improving the training. If people have requested copies of the materials you can create zip files from the CD and email them. If email is not available, it may be possible to follow up by letter or fax.

The trainer should use participants' feedback when designing subsequent workshops, and the follow-up email or letter is an opportunity to let participants know how their feedback has been used. For example: 'because of the suggestion made by several participants we will give more time for the Agree or Disagree exercise'.

If there were outstanding issues at the end of the workshop, make sure that you follow up on them by giving further references and information to participants. Tell participants it will help you if they tell you in the weeks and months after the workshop what use they have made of their learning. Let them know that further feedback and reflections on the training are useful at any time, and tell them how they can contact you. Share further feedback with trainers delivering the workshops in other countries.

# Improving the safety of civilians – feedback form

| Workshop location: | Date: |
|---|---|
| **Please tell us…** | |
| One thing you learned at the workshop | |
| What was your favourite part of the workshop and why? | |
| Were any parts difficult for you to understand? | |
| Any suggestions for improving the workshop next time? | |

| **After this workshop do you feel you…** | YES ☺ | | please tick ✓ ☺ | | NO ☹ |
|---|---|---|---|---|---|
| …understand what protection is? | | | | | |
| …know the standards for civilian protection? | | | | | |
| …know who the main protection actors are and their roles? | | | | | |
| …can do a protection analysis? | | | | | |
| …can plan a protection programme? | | | | | |
| …understand better what you can do when faced with a protection issue? | | | | | |
| …know how the protection core group works in your country? | | | | | |
| Other comments: | | | | | |

www.ingramcontent.com/pod-product-compliance
Lightning Source LLC
Chambersburg PA
CBHW060959030426
42334CB00033B/3291